Self-Improvement

Mastering the Art of Personal Growth

GALERON PRESS

GALERON CONSULTING

Copyright © 2023 by GALERON PRESS

All rights reserved.

No portion of this book may be reproduced in any form without written permission from the publisher or author, except as permitted by U.S. copyright law.

Contents

Dedication	VI
Epigraph	VII
Foreword	VIII
Introduction	IX
Prologue	X
1. Chapter 1: Unleashing Your True Potential	1
2. Identifying Your Core Values	2
3. Embracing Change and Overcoming Fears	14
4. Setting SMART Goals for Personal Growth	24
5. Building Resilience and Grit	37
6. Chapter 2: Developing a Success Mindset	49
7. Cultivating a Growth Mindset	50

8.	Embracing the Power of Positive Thinking	68
9.	Mastering Self-Discipline and Time Management	81
10.	Leveraging Visualization Techniques	92
11.	Chapter 3: Nurturing Relationships for Personal Success	105
12.	Enhancing Communication Skills	106
13.	Strengthening Emotional Intelligence	118
14.	Building and Maintaining Supportive Networks	128
15.	Conflict Resolution and Negotiation Skills	139
16.	Chapter 4: Achieving Balance in Life	149
17.	Managing Stress and Overcoming Burnout	150
18.	Incorporating Mindfulness and Meditation	161
19.	Establishing Work-Life Harmony	175
20.	Prioritizing Self-Care and Self-Compassion	183
21.	Chapter 5: Unstoppable Momentum: Continuously Evolving and Adapting	202
22.	Embracing Lifelong Learning	203
23.	Celebrating Progress and Overcoming Setbacks	215
24.	Developing Adaptability and Flexibility	228
25.	Staying Inspired and Motivated for Long-Term Success	240

Epilogue	250
Afterword	251
Acknowledgments	252

This book is dedicated to all those who are embarking on the journey of self-improvement. May you find the strength, courage, and perseverance to overcome obstacles and achieve your dreams.

"Change is the end result of all true learning." - Leo Buscaglia

Foreword

As an experienced life coach and motivational speaker, I have seen firsthand the transformative power of self-improvement. This book represents years of research, personal experience, and insights from countless individuals who have overcome challenges and achieved success. It is my hope that these pages will serve as a valuable resource for anyone seeking to improve their lives and reach their full potential.

INTRODUCTION

Welcome to the world of self-improvement. In this book, you will find a wealth of information, insights, and practical advice aimed at helping you become the best version of yourself. We will explore the importance of setting and achieving meaningful goals, cultivating healthy habits, developing emotional intelligence, and maintaining unstoppable momentum. By the end of this book, you will be equipped with the knowledge and tools necessary to create lasting change in your life.

Prologue

In today's fast-paced world, we are constantly seeking ways to improve ourselves, achieve our goals, and lead more fulfilling lives. This book is designed to serve as a comprehensive guide for those who are eager to embark on a journey of self-improvement and personal growth. Through actionable advice, practical tools, and inspirational stories, you will be empowered to take control of your life and strive for continuous evolution and adaptability.

Chapter 1: Unleashing Your True Potential

IDENTIFYING YOUR CORE VALUES

As a life coach, I've seen countless individuals embark on the journey to self-improvement. A crucial first step in this process is identifying your core values the guiding principles that shape your decisions, behavior, and life purpose. In this section, we'll explore various strategies to help you uncover your core values and align your life with them.

To begin, let's examine the importance of core values. They serve as a compass, guiding you through life's twists and turns. By identifying your core values, you're setting the foundation for a more fulfilling life, aligning your actions and decisions with what truly matters to you.

Now, let's dive into some practical steps to help you discover your core values:

Reflect on your life experiences: Consider the moments in your life when you felt most alive, proud, or fulfilled. What values were present during these experiences? Conversely, think about times when you felt frustrated, disappointed, or uneasy – what values were absent or compromised?

Create a list of potential values: Write down a list of values that resonate with you. These could include honesty, integrity, family, creativity, health, or financial security. Don't be concerned about narrowing down the list just yet – the goal is to brainstorm and consider what truly matters to you.

Prioritize your values: From your list, select the top five values that hold the most meaning for you. These are your core values. Remember, there are no right or wrong answers – your core values are deeply personal and unique to you.

Test your values: Reflect on whether your chosen values align with your daily actions and decisions. Are there areas in your life where you're not living in accordance with your core values? Identifying these discrepancies can help you make more intentional choices moving forward.

As an experienced motivational speaker, I can attest to the transformative power of understanding your core values. Let me share a real-life example to illustrate this point. Sarah, a client of mine, struggled with her career direction. She felt unfulfilled and lost, unable to pinpoint the source of her discontent. Through our sessions, Sarah discovered that her core values were creativity, connection, and personal growth. She realized her current job in finance didn't align with these values, which led her to pursue a new career as an art therapist – a role that allowed her to live in harmony with her core values.

To deepen your understanding of core values, I recommend the following resources:

Book: "The Values Factor" by Dr. John F. Demartini

Online course: "Discovering Your Core Values" by Brene Brown

Article: "7 Steps to Discovering Your Personal Core Values" by Scott Jeffrey

As you embark on your journey to self-improvement, take the time to answer the following questions:

What are your top five core values, and why do they hold importance for you?

How do your core values align with your current life choices and decisions?

In what areas of your life do you need to make changes to live more authentically with your core values?

By answering these questions and embracing your core values, you'll unlock your true potential and lay the foundation for lasting personal growth. Remember, as a life coach, consultant, and motivational speaker, I believe in your ability to create a life that aligns with your values and brings you immense joy and fulfillment.

As an experienced life coach and motivational speaker, I have encountered numerous individuals on their journey to personal growth. Here are three real-life examples that demonstrate the power of identifying core values in overcoming challenges and finding fulfillment:

Example 1: John's Struggle with Work-Life Balance

Problem: John, a successful entrepreneur, felt overwhelmed by his responsibilities and struggled to balance his work and personal life. He sought guidance on how to prioritize and find more fulfillment.

Solution: Through coaching, John discovered that his core values were family, integrity, and personal growth. With this newfound clarity, he made significant changes in his work approach, delegating more responsibilities, and setting clear boundaries. By aligning his life with his core values, John experienced improved work-life balance and increased satisfaction.

Example 2: Lisa's Search for Purpose

Problem: Lisa, a recent college graduate, was unsure of her career path and lacked a sense of purpose. She was overwhelmed by the possibilities and struggled to make decisions about her future.

Solution: As an experienced consultant, I guided Lisa through the process of identifying her core values. She realized that her top values were creativity, social impact, and learning. With this understanding, Lisa decided to pursue a career in non-profit management, where she could use her creativity to drive positive change and continue learning. Her newfound clarity and alignment with her core values led her to a fulfilling and purposeful career.

Example 3: Mark's Difficulty with Relationships

Problem: Mark faced recurring conflicts in his relationships, leaving him feeling disconnected and lonely. He sought guidance on how to improve his connections and build stronger bonds with others.

Solution: Through our sessions as a life coach, I helped Mark uncover his core values of honesty, empathy, and loyalty. We identified areas in his life where he wasn't living in alignment with these values, particularly in his communication style. By consciously incorporating his core values into his interactions, Mark improved his relationships, deepened his connections, and experienced greater satisfaction in his personal life.

In each of these examples, identifying core values played a crucial role in overcoming challenges and finding fulfillment. By understanding what truly matters to them, these individuals made intentional choices and aligned their lives with their values, leading to lasting personal growth.

Remember, as a life coach, consultant, and motivational speaker, I am here to guide you on your journey to unleashing your true potential. By identifying your core values and embracing them, you can overcome obstacles, make empowered decisions, and live a life that brings you joy and fulfillment.

As an experienced life coach and motivational speaker, I understand the importance of addressing common challenges related to identifying core values. Here are 20 questions and answers to help you overcome any obstacles and clarify your guiding principles:

Q: How do I begin the process of identifying my core values?

A: Start by reflecting on your life experiences, considering moments of joy, pride, frustration, and disappointment to identify values that were present or missing during those times.

Q: How many core values should I have?

A: Aim to identify your top five core values, as these will serve as your guiding principles in life.

Q: Can my core values change over time?

A: Yes, your core values may evolve as you grow and experience different stages in your life.

Q: How do I prioritize my core values?

A: Reflect on the values that hold the most meaning for you and consider which ones you cannot live without. These will be your top priorities.

Q: What if I'm having trouble narrowing down my list of core values?

A: Consider consulting with a life coach, therapist, or mentor to help you explore your values further and gain clarity.

Q: Can I have core values that conflict with each other?

A: It's possible to experience conflicts between your values. Identifying these conflicts can help you make more intentional choices and find balance.

Q: How do I align my life with my core values?

A: Begin by assessing your current choices and behaviors, and determine whether they are in harmony with your core values. Make necessary changes to live more authentically.

Q: Can my core values help me make better decisions?

A: Absolutely! Your core values serve as a compass, guiding your choices and actions in alignment with what truly matters to you.

Q: How can I stay true to my core values during challenging times?

A: Remind yourself of your values regularly, and consider how your decisions align with them. Reflect on the long-term consequences of your choices and stay committed to your principles.

Q: Should I share my core values with others?

A: Sharing your core values with those close to you can deepen your connections and help you stay accountable to your principles.

Q: How do I know if my core values are healthy and positive?

A: Healthy core values are those that contribute to your well-being, growth, and positive relationships with others. Reflect on whether your values are helping or hindering your progress.

Q: How can I incorporate my core values into my daily life?

A: Set intentions, create habits, and establish boundaries that support your core values, making them an integral part of your everyday life.

Q: Can my core values help me achieve my goals?

A: Yes, by aligning your goals with your core values, you can create more meaningful objectives that resonate with your true self.

Q: How do I know if I'm living in alignment with my core values?

A: Assess your choices, behaviors, and overall life satisfaction. If you find harmony and contentment, it's likely that you are living in alignment with your values.

Q: What if my core values differ from those around me?

A: Embrace your unique values and respect the differences of others. Engaging in open, respectful dialogue can foster understanding and acceptance.

Q: How do I handle situations that challenge my core values?

A: Stand firm in your convictions, communicate your values respectfully, and seek support from those who share your principles.

Q: What if my work environment conflicts with my core values?

A: Consider discussing your concerns with your supervisor or HR department to explore potential solutions. If the conflict persists, it may be time to seek a new opportunity that aligns with your values.

Q: How can I ensure my relationships align with my core values?

A: Communicate openly about your values with your loved ones and seek relationships that support and respect your principles.

Q: Can identifying my core values help me overcome personal challenges?

A: Yes, by understanding your core values, you can approach challenges with a clearer perspective and make choices that align with your true self.

Q: How can I continue to grow and develop my understanding of my core values?

A: Engage in regular self-reflection, seek feedback from trusted individuals, and consider working with a life coach or therapist to gain deeper insights into your values.

By addressing these questions and embracing your core values, you'll be well on your way to unleashing your true potential. As a life coach, consultant, and motivational speaker, I'm confident in your ability to create a life that aligns with your values, bringing you joy, fulfillment, and personal growth. Stay true to your principles and trust your journey as you strive to become the best version of yourself.

As an experienced life coach, consultant, and motivational speaker, I understand the importance of providing practical tools and techniques for readers to apply the concepts discussed in this book. Here are six techniques that you can follow to help you identify and embody your core values:

Technique 1: Core Values Reflection Exercise

Step 1: Reflect on moments in your life when you felt most proud, fulfilled, or joyful. Jot down these experiences.

Step 2: Identify the values that were present in those moments. List them out.

Step 3: Reflect on times when you felt frustrated, disappointed, or upset. Identify the values that were absent or compromised during these moments.

Technique 2: Values Ranking Exercise

Step 1: Compile a list of potential core values (15-20 values) that resonate with you.

Step 2: Narrow down the list by ranking the values in order of importance, eliminating values that do not align with your true self.

Step 3: Select your top five core values to serve as your guiding principles.

Technique 3: Values-Based Decision-Making Worksheet

Step 1: When faced with a challenging decision, list the options available to you.

Step 2: Consider how each option aligns with your core values.

Step 3: Assess the long-term consequences of each choice and select the option that best aligns with your values.

Technique 4: Daily Core Values Alignment Check-In

Step 1: Create a simple chart or journal page with your top five core values.

Step 2: At the end of each day, reflect on how your actions and choices aligned with your values.

Step 3: Identify areas for improvement and set intentions for the following day to better embody your values.

Technique 5: Core Values Vision Board

Step 1: Collect images, quotes, and symbols that represent your core values.

Step 2: Arrange the items on a board, poster, or digital platform.

Step 3: Display your vision board in a prominent location as a daily reminder of your guiding principles.

Technique 6: Guided Meditation for Embodying Core Values

Step 1: Find a quiet space and sit comfortably, closing your eyes and taking a few deep breaths.

Step 2: Visualize each of your core values, imagining scenarios where you embody these values fully.

Step 3: Reflect on the emotions and sensations that arise while embodying your values. Allow these feelings to inspire and motivate you.

By following these techniques and incorporating them into your daily life, you'll be well on your way to unleashing your true potential. As a life coach, consultant, and motivational speaker, I'm confident in your ability to live authentically and create a life that aligns with your values, bringing you joy, fulfillment, and personal growth. Embrace your journey and stay committed to your principles as you strive to become the best version of yourself.

Embracing Change and Overcoming Fears

Change is an inevitable part of life, and overcoming fears associated with change is essential to unlocking your true potential. As a life coach, consultant, and motivational speaker, I understand the importance of embracing change to grow and thrive personally and professionally. In this section, we'll explore strategies for embracing change and overcoming fears that hold you back from becoming your best self.

Understanding the Nature of Change

Recognize that change is constant and necessary for growth. Change can be uncomfortable, but it's through this discomfort that we learn, adapt, and evolve. Embracing change is essential to personal development and achieving your goals.

Overcoming Fear of the Unknown

Fear of the unknown can be paralyzing, preventing you from moving forward and embracing new opportunities. To overcome this fear, focus on what you can control, gather information about the potential change, and make informed decisions.

Building Resilience

Resilience is the ability to bounce back from setbacks and adapt to change. Develop your resilience by practicing self-awareness, managing your emotions, and cultivating a positive mindset.

Taking Small Steps

Break down change into manageable steps to make it less overwhelming. Gradually taking small steps towards change can help you build confidence and reduce fear.

Developing a Support Network

Surround yourself with supportive and like-minded people who can help you navigate change and provide encouragement when you face challenges.

Practicing Self-Compassion

Acknowledge that change can be challenging and that it's normal to feel fear. Practice self-compassion and treat yourself kindly as you work through these emotions.

Suggested Readings:

"Who Moved My Cheese?" by Spencer Johnson

"The Power of Now: A Guide to Spiritual Enlightenment" by Eckhart Tolle

"Daring Greatly: How the Courage to Be Vulnerable Transforms the Way We Live, Love, Parent, and Lead" by Brené Brown

Real-Life Example:

Julia, a successful marketing executive, faced a significant change in her career when her company underwent restructuring. Despite her fear of the unknown, she used the strategies above to embrace the change and adapt to her new role. By gathering information about the restructuring process, taking small steps to transition, and relying on her support network, Julia overcame her fears and ultimately thrived in her new position.

Question:

How can I apply these strategies to overcome my fear of change and embrace new opportunities?

Answer:

Identify a specific change or challenge in your life and apply the strategies discussed in this section. Recognize the nature of change, gather information, build resilience, take small steps, develop a support network, and practice self-compassion as you navigate the change.

By embracing change and overcoming fears, you'll be one step closer to unleashing your true potential. As a life coach, consultant, and motivational speaker, I'm confident that you can transform your life by incorporating these strategies and living authentically. Stay committed to your personal growth and trust your journey as you strive to become the best version of yourself.

Real-Life Story 1:

Problem: Sarah, a dedicated employee at a tech company, was offered a promotion that required her to relocate to another city. Although excited about the opportunity, she was hesitant to embrace the change due to her fear of leaving her support system and starting a new life in an unfamiliar place.

Solution: As a life coach, I encouraged Sarah to gather information about the new city, explore the benefits of the promotion, and develop a plan for maintaining her support system despite the distance. By breaking down the change into manageable steps and focusing on the positives, Sarah was able to overcome her fears, accept the promotion, and thrive in her new role and city.

Real-Life Story 2:

Problem: Mark, a middle-aged man, faced a significant life change when he decided to switch careers after years in the same industry. He struggled with fear and uncertainty, questioning whether he made the right decision and if he could succeed in a new field.

Solution: As a motivational speaker, I shared strategies for embracing change and overcoming fears with Mark. He began to practice self-compassion, acknowledging that it's normal to feel fear during significant life changes. Mark focused on building his resilience by actively seeking new learning opportunities and cultivating a positive mindset. Through these efforts, he successfully transitioned into his new career and found fulfillment in his work.

Real-Life Story 3:

Problem: Emma, a recent college graduate, was anxious about entering the workforce and leaving the familiarity of her academic environment. She worried about the challenges she would face and whether she could adapt to her new professional life.

Solution: As an experienced consultant, I worked with Emma to create a plan for her transition into the workforce. We identified her core values and set achievable goals to help her stay focused on her personal and professional growth. By developing a support network of mentors and peers in her field, Emma gained the confidence to embrace the change and overcome her fears, ultimately excelling in her new career.

By learning from these real-life stories, you can apply the strategies shared in this chapter to your own life, embracing change and overcoming fears that hold you back from reaching your true potential. As a life coach, consultant, and motivational speaker, I'm confident that you can transform your life by incorporating these strategies and living authentically. Stay committed to your personal growth and trust your journey as you strive to become the best version of yourself.

Question: How can I identify the fears that are holding me back from embracing change?

Answer: Reflect on your thoughts and emotions when faced with change, and journal your feelings to uncover any underlying fears. You can also consult with a trusted friend, mentor, or life coach to gain insight into your fears.

Question: What steps can I take to overcome my fear of the unknown?

Answer: Gather information about the potential change, focus on what you can control, and take small, manageable steps toward the change to build your confidence.

Question: How can I develop resilience in the face of change?

Answer: Practice self-awareness, manage your emotions, and cultivate a positive mindset. Embrace challenges as opportunities for growth and learn from your experiences.

Question: How do I build a support network to help me navigate change?

Answer: Reach out to friends, family, mentors, or colleagues who share your values and goals. Join clubs, professional organizations, or online forums to connect with like-minded individuals.

Question: How can I practice self-compassion when facing change and fear?

Answer: Acknowledge that it's normal to feel fear during change and treat yourself kindly. Focus on your strengths and accomplishments, and remember that growth often involves discomfort.

Question: How can I break down a significant change into manageable steps?

Answer: Identify the main components of the change and create a step-by-step plan to tackle each component. Set achievable goals and celebrate your progress along the way.

Question: How can I maintain a positive mindset during times of change?

Answer: Focus on the opportunities and growth that can come from change, surround yourself with positive influences, and practice gratitude for the good things in your life.

Question: How can I use my core values to guide my decision-making during times of change?

Answer: Reflect on your core values and consider how they align with the potential change. Let your values guide your choices and actions, ensuring that the change aligns with your authentic self.

Question: How can I develop my problem-solving skills to better adapt to change?

Answer: Practice critical thinking, brainstorming, and reflection. Seek out new learning opportunities and challenge yourself to find creative solutions to problems.

Question: How can I overcome the fear of failure during change?

Answer: Embrace the possibility of failure as a learning opportunity. Remember that setbacks can provide valuable lessons and help you grow.

Question: How can I manage stress during periods of change?

Answer: Develop a self-care routine that includes exercise, proper nutrition, and adequate sleep. Practice mindfulness techniques, such as meditation or deep breathing, to stay centered and focused.

Question: How can I maintain my motivation during challenging times of change?

Answer: Set clear, achievable goals and track your progress. Celebrate your successes and remind yourself of your core values and long-term vision.

Question: How can I cultivate a growth mindset to embrace change more easily?

Answer: View challenges as opportunities for growth and development. Emphasize effort, persistence, and continuous learning rather than focusing solely on outcomes.

Question: How can I communicate my needs and boundaries to others during times of change?

Answer: Be open and honest about your feelings, needs, and boundaries. Seek support and understanding from those around you, while also respecting their perspectives and boundaries.

Question: How can I maintain a work-life balance during significant changes?

Answer: Prioritize self-care and set boundaries to protect your personal time. Communicate your needs to your employer or colleagues, and seek support from your network.

Question: How can I develop my emotional intelligence to better manage change and fear?

Answer: Practice self-awareness, empathy, and effective communication. Reflect on your emotions and seek to understand the emotions of others. Engage in active listening and express your feelings in a healthy, constructive manner.

Question: How can I recognize when it's time to seek professional help to overcome my fears or navigate change?

Answer: If your fears or resistance to change are causing significant distress or impacting your daily functioning, it may be time to seek the help of a therapist, life coach, or counselor.

Question: How can I stay focused on my goals during times of change?

Answer: Regularly review your goals and assess your progress. Break down your goals into smaller, achievable milestones and create a plan to stay on track. Stay accountable by sharing your goals with a trusted friend or mentor.

Question: How can I embrace change and overcome fears when making a major career shift?

Answer: Research your new career path, identify transferable skills, and network with professionals in the field. Develop a plan to build necessary skills and gain experience, and seek support from mentors and peers to navigate the transition.

Question: How can I handle change in my personal relationships while still prioritizing my own growth and well-being?

Answer: Communicate openly and honestly with your loved ones, expressing your needs and boundaries. Prioritize self-care, maintain a support network, and be open to renegotiating the dynamics of your relationships as you grow and change.

By exploring these questions and answers, you can develop a deeper understanding of how to embrace change and overcome fears. As you continue to work through this book, remember that growth often requires facing challenges and discomfort. By cultivating resilience, embracing a growth mindset, and staying true to your core values, you can unleash your true potential and thrive in times of change.

Setting SMART Goals for Personal Growth

Setting goals is an essential aspect of personal growth and achieving success. However, merely having goals is not enough; they need to be SMART – Specific, Measurable, Achievable, Relevant, and Time-bound. By creating SMART goals, you'll be better equipped to track your progress, stay motivated, and ultimately achieve the results you desire.

Specific: Clearly define your goals, outlining the steps you need to take and the results you hope to achieve. A specific goal answers the questions: What do I want to accomplish? Why is it important? How will I achieve it?

Example: Instead of "I want to improve my public speaking skills," a specific goal could be, "I will join a local Toastmasters club and attend weekly meetings to improve my public speaking skills."

Measurable: Establish criteria for measuring your progress and success. A measurable goal answers the question: How will I know when I've achieved my goal?

Example: A measurable goal could be, "I will deliver five speeches at Toastmasters meetings, receive feedback from fellow members, and refine my skills based on that feedback."

Achievable: Set realistic and attainable goals that challenge you but are still within reach. An achievable goal answers the question: Can I realistically accomplish this goal given my current resources, skills, and constraints?

Example: If you've never given a speech before, setting a goal to speak at a TEDx event within a month might not be achievable. A more attainable goal could be, "I will work on my public speaking skills for six months and apply to speak at a local TEDx event."

Relevant: Align your goals with your values, priorities, and long-term objectives. A relevant goal answers the question: Does this goal support my broader vision and purpose?

Example: If your long-term goal is to become a motivational speaker, improving your public speaking skills through Toastmasters is a relevant goal.

Time-bound: Set a deadline for achieving your goals to create a sense of urgency and stay focused. A time-bound goal answers the question: When will I achieve this goal?

Example: A time-bound goal could be, "I will improve my public speaking skills and apply to speak at a local TEDx event within the next year."

In addition to setting SMART goals, it's crucial to monitor your progress regularly, adjust your goals as needed, and celebrate your achievements along the way.

Suggested Readings:

Locke, E. A., & Latham, G. P. (2006). New Directions in Goal-Setting Theory. Current Directions in Psychological Science, 15(5), 265–268.

Doran, G. T. (1981). There's a S.M.A.R.T. Way to Write Management's Goals and Objectives. Management Review, 70(11), 35-36.

Real-life Example:

Jane, a marketing professional, wanted to advance her career and become a marketing director. She set a SMART goal: "I will complete a digital marketing certification course within the next six months to improve my skills and increase my chances of being promoted to marketing director." Jane made a plan, followed through, and achieved her goal within the specified timeframe. As a result, she was able to secure the promotion she desired.

Questions & Answers:

Q: How can I ensure my goals are specific enough?

A: Make sure your goals answer the "what, why, and how" questions. Break them down into smaller, actionable steps to make them more specific.

Q: How do I know if my goal is achievable?

A: Assess your current resources, skills, and constraints. If the goal is challenging but still within your capabilities, it's likely achievable. You can also seek guidance from mentors or peers who have achieved similar goals to gauge whether your goal is realistic.

Q: How can I make my goals more relevant?

A: Evaluate your goals in the context of your long-term vision and purpose. If the goal aligns with your broader objectives and values, it's likely relevant.

Q: What if I don't achieve my goal within the set timeframe?

A: It's important to be flexible and adjust your goals when necessary. If you don't meet your deadline, reassess your goal, identify any obstacles or challenges that hindered your progress, and adjust your plan or timeline accordingly.

Q: How can I stay motivated while working towards my goals?

A: Break your goals into smaller, manageable milestones, and celebrate your achievements along the way. Regularly review your progress and remind yourself of the reasons behind your goals to stay motivated.

Chapter 1: Unleashing Your True Potential

Section 1.3: Setting SMART Goals for Personal Growth —.

Technique 1: Goal-setting Worksheet

Create a goal-setting worksheet that includes columns for each aspect of the SMART criteria. Fill in the worksheet with your goals, ensuring each one meets the SMART criteria. Review and update the worksheet regularly to track your progress.

Technique 2: Visualization

Regularly visualize yourself achieving your goals, imagining the emotions and rewards associated with success. Visualization can help reinforce your commitment to your goals and enhance motivation.

Technique 3: Accountability Partner

Find an accountability partner who can help you stay on track with your goals. Share your SMART goals with them, and schedule regular check-ins to discuss your progress, challenges, and any adjustments needed.

Technique 4: Progress Journal

Maintain a progress journal to document your journey towards your goals. Record your achievements, challenges, and insights, and reflect on your growth and development.

Technique 5: Reward System

Establish a reward system for achieving milestones along your goal journey. Treat yourself to something enjoyable, like a spa day or a special meal, to celebrate your progress and maintain motivation.

By following these techniques and tools, you'll be better equipped to set and achieve SMART goals, paving the way for personal growth and success.

Example 1: Overcoming Procrastination with SMART Goals

Problem: Janet, a 35-year-old freelance graphic designer, struggles with procrastination. She often waits until the last minute to complete projects, which causes her stress and negatively impacts the quality of her work.

Solution: Janet's life coach suggests setting a SMART goal to overcome her procrastination. Together, they create the goal: "Complete all client projects at least two days before the deadline for the next three months." This goal is Specific (completing projects), Measurable (two days before the deadline), Achievable (Janet has control over her schedule), Relevant (it directly addresses her procrastination problem), and Time-bound (three months).

Result: Janet diligently works towards her goal and soon notices a decrease in stress levels and an improvement in her work quality. By setting and achieving her SMART goal, she gains better control over her procrastination habit.

Example 2: Balancing Work and Family Life

Problem: Tom, a 45-year-old sales manager, struggles to balance his demanding job with spending quality time with his family. He often works long hours, causing him to miss important family events and strain his relationships.

Solution: Tom's motivational speaker advises him to set a SMART goal to improve his work-life balance. He decides on the following goal: "Spend at least two evenings per week and one full weekend day with my family, without discussing work, for the next six months." This goal is Specific (spending time with family), Measurable (two evenings and one weekend day), Achievable (Tom can adjust his work schedule), Relevant (it addresses his work-life balance), and Time-bound (six months).

Result: By setting and following this SMART goal, Tom is able to strengthen his relationships with his family and improve his overall well-being, all while maintaining his job performance.

Example 3: Enhancing Public Speaking Skills

Problem: Maria, a 28-year-old marketing professional, experiences anxiety when presenting in front of large groups. She wants to overcome her fear and become a more effective public speaker.

Solution: Maria's experienced consultant recommends setting a SMART goal to improve her public speaking skills. They work together to create the goal: "Attend a public speaking workshop and deliver three presentations in front of at least 30 people within the next four months." This goal is Specific (attending a workshop and delivering presentations), Measurable (three presentations), Achievable (Maria can sign up for a workshop and seek speaking opportunities),

Relevant (it directly addresses her public speaking anxiety), and Time-bound (four months).

Result: Maria follows her SMART goal, attends the workshop, and practices her new skills in front of large groups. Over time, her confidence grows, and she becomes a more effective and composed public speaker.

Q: What does SMART stand for in goal setting?

A: SMART stands for Specific, Measurable, Achievable, Relevant, and Time-bound. These criteria help create clear, focused, and attainable goals.

Q: How do SMART goals contribute to personal growth?

A: SMART goals promote personal growth by providing a clear path, structure, and timeline to achieve objectives, making it easier to stay motivated and track progress.

Q: Can SMART goals be applied to both short-term and long-term goals?

A: Yes, SMART goals can be used for both short-term and long-term goals. The time-bound aspect helps determine the appropriate timeframe for each goal.

Q: How can I make my goals more specific?

A: To make goals more specific, focus on the desired outcome, and include details about what needs to be accomplished, when, and how.

Q: What does it mean for a goal to be measurable?

A: A measurable goal includes specific criteria or indicators that allow you to track progress and determine when the goal has been achieved.

Q: How do I know if my goal is achievable?

A: A goal is achievable if it is realistic, considering your resources, skills, and constraints, and it pushes you to grow without being unattainable.

Q: Why is it important for a goal to be relevant?

A: A relevant goal aligns with your values, priorities, and long-term vision, ensuring that the effort you put into achieving it contributes to your overall personal growth.

Q: How can I set a time-bound goal if I'm unsure how long it will take to achieve?

A: Estimate a reasonable timeframe based on similar past experiences, research, or by seeking advice from others who have achieved similar goals. Adjust the timeline as needed while pursuing the goal.

Q: How often should I review and adjust my SMART goals?

A: Review your SMART goals regularly, ideally monthly or quarterly, to assess progress, address any obstacles, and make necessary adjustments to stay on track.

Q: Can I work on multiple SMART goals simultaneously?

A: Yes, you can work on multiple goals, but it's essential to prioritize and ensure that you have the time, energy, and resources to commit to each goal.

Q: What should I do if I'm not making progress toward my SMART goal?

A: If you're not making progress, reassess your approach, identify obstacles, and adjust your goal or the steps needed to achieve it. Seek support or advice from a life coach, mentor, or others who have achieved similar goals.

Q: How can I stay motivated while working towards my SMART goals?

A: Stay motivated by regularly reviewing your goals, celebrating small milestones, seeking support from others, and reminding yourself of the long-term benefits of achieving the goal.

Q: What if I achieve my SMART goal earlier than expected?

A: If you achieve your goal early, celebrate your success, then consider setting new goals or revisiting other areas of personal growth to continue your journey.

Q: How can I overcome obstacles or setbacks when pursuing my SMART goals?

A: Overcome obstacles by analyzing the situation, adapting your plan, seeking support, and focusing on the long-term benefits of achieving your goal.

Q: How can I track my progress towards achieving my SMART goals?

A: Track progress by using a journal, spreadsheet, or app to document milestones, challenges, and accomplishments related to your goals.

Q: Can I adjust my SMART goals if my priorities change?

A: Yes, adjust your goals as needed to align with your evolving priorities and ensure they

Q: How can I ensure my SMART goals are in line with my core values?

A: Reflect on your core values and assess whether your goals align with them. If they don't, consider revising your goals to better match your values and priorities.

Q: Can SMART goals help improve my time management and productivity?

A: Yes, SMART goals can improve time management and productivity by providing a clear structure, timeline, and focus for your efforts, making it easier to prioritize tasks and stay on track.

Q: How can I stay accountable while working towards my SMART goals?

A: Stay accountable by sharing your goals with a trusted friend, family member, or mentor, seeking their support and feedback, and regularly reporting on your progress.

Q: What if I don't achieve my SMART goal within the specified timeframe?

A: If you don't achieve your goal within the timeframe, assess the reasons for this, learn from the experience, and adjust your goal or the steps needed to achieve it. Remember, setbacks are a natural part of the personal growth process and can provide valuable learning opportunities.

Create a goal-setting worksheet to help you outline and track your SMART goals. This worksheet should include sections for:

Goal description

Specificity

Measurability

Achievability

Relevance

Time-bound

Regularly review and update this worksheet as you progress, adjusting goals as needed.

Technique 2: Visualization Exercises

Practice visualization exercises to imagine the desired outcome of achieving your SMART goals. Visualization can help you stay motivated, focused, and inspired. Spend a few minutes daily visualizing your success in detail.

Technique 3: Break Goals into Milestones

Break your larger SMART goals into smaller milestones. These milestones will serve as stepping stones toward your ultimate goal, making it more manageable and easier to monitor progress.

Technique 4: Prioritize and Organize with a Goal Planner

Use a goal planner to prioritize and organize your daily, weekly, and monthly tasks related to your SMART goals. A planner will help you stay focused and manage your time effectively.

Technique 5: Create a Support Network

Share your SMART goals with a trusted group of friends, family members, or mentors who can provide encouragement, guidance, and accountability. Regularly update them on your progress and ask for feedback.

Technique 6: Reward Yourself for Progress

Recognize and reward yourself when you achieve milestones or make significant progress toward your SMART goals. Celebrating achievements helps maintain motivation and reinforces positive behavior.

Technique 7: Reflect and Adjust

Periodically review your SMART goals and assess your progress. Reflect on your achievements, setbacks, and lessons learned. Use this insight to adjust your goals or strategies as necessary.

Incorporate these techniques into your personal growth journey to effectively set and achieve SMART goals. With dedication, persistence, and a growth mindset, you will be well on your way to unleashing your true potential.

BUILDING RESILIENCE AND GRIT

Resilience and grit are essential characteristics that enable individuals to adapt to challenges, overcome setbacks, and persevere in the pursuit of their goals. By cultivating these qualities, you can unlock your true potential and achieve personal growth.

Resilience is the ability to bounce back from adversity and maintain a positive outlook despite challenges. Grit, on the other hand, is the unwavering determination and commitment to achieve long-term goals, even in the face of obstacles. Both of these traits are vital for personal development and success.

The following strategies can help you build resilience and grit:

Adopt a growth mindset: Embrace the idea that your abilities and intelligence can be developed through hard work, persistence, and dedication. This mindset encourages you to view challenges as opportunities to learn and grow, rather than threats to your self-worth.

Develop strong coping skills: Learn healthy ways to manage stress, such as mindfulness meditation, regular exercise, and maintaining a balanced diet. Practicing good self-care can help you stay resilient during difficult times.

Cultivate a support network: Surround yourself with positive influences and supportive relationships that encourage your growth and resilience. Reach out to friends, family, mentors, or support groups to share your experiences and seek guidance.

Set realistic expectations: Understand that setbacks are a natural part of personal growth and that you may not always achieve your goals on the first try. Be patient with yourself and maintain a realistic perspective on your progress.

Reflect on past experiences: Consider the challenges you've faced in the past and identify the lessons learned from those experiences. By recognizing your ability to overcome adversity, you can build confidence in your resilience and grit.

Suggested readings for building resilience and grit include:

"Grit: The Power of Passion and Perseverance" by Angela Duckworth

"Resilience: Hard-Won Wisdom for Living a Better Life" by Eric Greitens

"Mindset: The New Psychology of Success" by Carol S. Dweck

Real-life example:

Jane, a marketing executive, faced a series of setbacks at work, including project failures and harsh criticism from her manager. Instead of giving up, she adopted a growth mindset, viewing these challenges as opportunities to learn and improve. Jane sought feedback from colleagues and mentors, implemented new strategies, and ultimately earned a promotion for her perseverance and dedication.

Question: How can I develop a growth mindset?

Answer: Focus on learning from your experiences, embrace challenges as opportunities for growth, and believe that your abilities can improve with effort and persistence.

By incorporating these strategies and focusing on personal growth, you can build resilience and grit, unlocking your true potential and achieving your goals. Remember, as a life coach, motivational speaker, or experienced consultant, your guidance and encouragement can make a significant difference in the lives of those you inspire.

Real-life story 1:

Tom, a small business owner, was forced to close his store temporarily due to unforeseen circumstances. He felt devastated, but instead of giving up, he decided to take this time as an opportunity to reflect on his business strategy.

Tom reached out to an experienced consultant who helped him identify areas for improvement. With the consultant's guidance, Tom revamped his business plan, strengthened his online presence, and reopened his store with renewed resilience and grit. Now, his business is thriving more than ever before.

Problem/Question: How can I bounce back from a setback in my business?

Answer: Reach out to experienced professionals, reevaluate your business strategy, and focus on areas for improvement. Embrace challenges as opportunities for growth and develop resilience and grit.

Real-life story 2:

Samantha, a young athlete, faced a severe injury that prevented her from participating in her sport for several months. Instead of allowing the setback to define her, Samantha chose to use this time to develop mental toughness and resilience. As a motivational speaker, she began sharing her journey of overcoming adversity with others, inspiring them to embrace challenges and cultivate grit. Samantha eventually returned to her sport, stronger and more determined than ever.

Problem/Question: How can I turn my setbacks into opportunities for personal growth?

Answer: Shift your perspective and view setbacks as opportunities to learn and grow. Use the experience to develop resilience and grit, and share your story to inspire and motivate others.

Real-life story 3:

Maria, a single mother, faced financial difficulties and struggled to make ends meet. She felt overwhelmed by her situation and uncertain about the future. Seeking guidance from a life coach, Maria learned the importance of resilience and grit in overcoming adversity. She set SMART goals to improve her financial situation and dedicated herself to achieving them. Over time, Maria's resilience and determination helped her overcome her challenges, and she began a successful career in real estate.

Problem/Question: How can I develop resilience and grit to overcome financial challenges?

Answer: Seek guidance from a life coach or experienced consultant, set SMART goals for personal growth, and commit to making the necessary changes in your life. Cultivate resilience and grit through determination and a growth mindset.

Question: How can I develop resilience in the face of adversity?

Answer: Embrace challenges as opportunities for growth, maintain a positive attitude, and learn from past experiences. Surround yourself with supportive people and practice self-care.

Question: What are some strategies for building grit in my personal and professional life?

Answer: Set long-term goals, practice perseverance, develop a strong work ethic, and cultivate a growth mindset. Embrace failure as a learning opportunity.

Question: How can I bounce back from a failure or setback?

Answer: Reflect on the experience, identify lessons learned, set new goals, and maintain a positive outlook. Use your newfound knowledge to make better decisions moving forward.

Question: How can resilience and grit help me in achieving my goals?

Answer: Resilience and grit can help you stay focused, overcome obstacles, and persist in the face of challenges, ultimately increasing your chances of success.

Question: What role does self-awareness play in building resilience and grit?

Answer: Self-awareness helps you understand your strengths and weaknesses, enabling you to develop strategies for overcoming challenges and cultivating resilience and grit.

Question: How can I develop mental toughness?

Answer: Practice mindfulness, maintain a healthy lifestyle, seek support from friends and mentors, and embrace challenges as opportunities for growth.

Question: Can resilience and grit be learned, or are they innate qualities?

Answer: While some individuals may have a natural inclination toward resilience and grit, these traits can be developed and strengthened through practice and experience.

Question: How can I help my children develop resilience and grit?

Answer: Encourage them to face challenges, teach them the value of perseverance, and provide a supportive environment that fosters growth and learning.

Question: How can I maintain motivation when faced with setbacks and obstacles?

Answer: Focus on your long-term goals, practice self-compassion, and remind yourself of past successes. Seek inspiration from others who have overcome similar challenges.

Question: What are some daily habits that can help me build resilience and grit?

Answer: Practice mindfulness, set small achievable goals, maintain a positive attitude, and engage in activities that build your self-confidence.

Question: How can I measure my progress in building resilience and grit?

Answer: Reflect on personal growth, track your achievements, and monitor improvements in your ability to overcome challenges and persist in the face of adversity.

Question: What role does having a support system play in building resilience and grit?

Answer: A strong support system provides encouragement, guidance, and accountability, helping you stay focused and committed to your goals.

Question: Can building resilience and grit help reduce stress and anxiety?

Answer: Yes, resilience and grit can help you better cope with stress and anxiety by fostering a sense of control and confidence in your ability to overcome challenges.

Question: How can I develop resilience and grit in the workplace?

Answer: Set achievable goals, embrace challenges, seek feedback, and develop a strong work ethic. Cultivate a growth mindset and learn from setbacks.

Question: How can I apply the concepts of resilience and grit to my relationships?

Answer: Practice open communication, show empathy, maintain a positive outlook, and be willing to work through challenges and conflicts together.

Question: How can I overcome the fear of failure while building resilience and grit?

Answer: Reframe failure as a learning opportunity, set realistic expectations, and focus on the process rather than the outcome.

Question: What are some resources for further learning about resilience and grit?

Answer: Books, podcasts, online courses, and workshops by experienced life coaches, consultants, and motivational speakers can provide valuable insights and guidance

Question: How can practicing self-compassion contribute to building resilience and grit?

Answer: Self-compassion helps you maintain a positive mindset, acknowledge your struggles, and treat yourself with kindness, making it easier to recover from setbacks and stay committed to your goals.

Question: How does setting boundaries contribute to resilience and grit?

Answer: Setting boundaries allows you to maintain a healthy balance in your life, ensuring you have the energy and focus needed to face challenges and work towards your goals.

Question: Can physical exercise help in building resilience and grit?

Answer: Yes, physical exercise can improve mental toughness, boost self-confidence, and promote overall well-being, all of which contribute to building resilience and grit.

By exploring these questions and their answers, you can gain a deeper understanding of resilience and grit and how they can help you unleash your true potential. Keep in mind that cultivating these qualities takes time and effort, but with practice and determination, you'll be well-equipped to overcome any challenges you may face on your journey to personal growth.

Technique 1: Develop a Growth Mindset

As a motivational speaker, I encourage you to embrace a growth mindset by viewing challenges as opportunities to learn and grow. To develop this mindset, practice the following steps:

Recognize your limiting beliefs and self-talk.

Replace negative thoughts with positive affirmations.

Embrace failures as learning experiences.

Continuously seek new opportunities for growth and learning.

Technique 2: Practice Mindfulness Meditation

Mindfulness meditation can help you build resilience and grit by increasing your self-awareness and ability to cope with stress. Follow these steps to start your mindfulness practice:

Find a quiet space and sit comfortably.

Close your eyes and focus on your breath.

Acknowledge any thoughts or emotions that arise, and let them pass without judgment.

Consistently practice mindfulness meditation for at least 10 minutes a day.

Technique 3: Create a Resilience Journal

Documenting your journey can provide valuable insights and motivate you to continue building resilience and grit. Follow these steps to start your resilience journal:

Dedicate a notebook or digital document for your journal.

Write down your thoughts, emotions, and experiences regularly.

Reflect on your progress and identify areas for improvement.

Celebrate your successes and learn from your setbacks.

Technique 4: Cultivate a Support Network

Having a strong support network can significantly impact your ability to build resilience and grit. To create a supportive community, follow these steps:

Reach out to friends, family, or colleagues who share similar goals or values.

Attend local events, workshops, or seminars to meet like-minded individuals.

Join online forums or social media groups related to your interests or goals.

Offer support and encouragement to others, fostering a mutually beneficial support system.

Technique 5: Develop Healthy Coping Strategies

Learning how to effectively manage stress and adversity is crucial to building resilience and grit. Here are some healthy coping strategies to consider:

Engage in regular physical activity to boost your mood and energy levels.

Practice deep breathing exercises or progressive muscle relaxation to reduce stress.

Utilize creative outlets such as painting, writing, or playing music to express your emotions.

Seek professional help from a life coach, therapist, or counselor if needed.

By incorporating these techniques and tools into your daily routine, you'll be well on your way to building resilience and grit, ultimately unleashing your true potential.

Chapter 2: Developing a Success Mindset

CULTIVATING A GROWTH MINDSET

Cultivating a growth mindset is fundamental to achieving success in any aspect of life. By believing in your ability to learn and grow, you open yourself up to new opportunities and experiences. In this section, we will explore the concept of a growth mindset, the difference between a growth mindset and a fixed mindset, and practical strategies to develop a growth mindset.

Understanding a Growth Mindset

A growth mindset is a belief that intelligence, abilities, and talents can be developed over time through dedication and hard work. This mindset allows individuals to embrace challenges, persevere through setbacks, and constantly seek self-improvement. Dr. Carol Dweck, a leading psychologist and researcher, first introduced the concept in her book "Mindset: The New Psychology of Success."

Growth Mindset vs. Fixed Mindset

In contrast to a growth mindset, a fixed mindset is the belief that intelligence and abilities are static and cannot be changed. Individuals with a fixed mindset often avoid challenges, give up easily, and view effort as fruitless. The difference between these two mindsets is crucial to understanding why some people achieve success while others struggle.

Strategies to Cultivate a Growth Mindset

As a life coach, I encourage you to embrace the following strategies to develop a growth mindset:

Reflect on your beliefs: Identify areas where you may hold a fixed mindset and challenge those beliefs. Understand that your intelligence and abilities are not fixed and can be improved with effort.

Embrace challenges: Instead of avoiding challenges, view them as opportunities to learn and grow. Push yourself out of your comfort zone and take calculated risks to expand your capabilities.

Focus on effort and progress: Celebrate the effort you put into achieving your goals, rather than solely focusing on the outcomes. Recognize that progress, no matter how small, is a step towards success.

Learn from failure: Rather than viewing failure as a dead-end, analyze your setbacks to identify areas for improvement. Use these lessons as stepping stones to propel you towards your goals.

Seek feedback: Regularly solicit constructive feedback from peers, mentors, or coaches. Use this feedback to refine your approach and develop new strategies for success.

Surround yourself with growth-minded individuals: Associate with people who share your belief in personal growth and development. Their support and encouragement will help reinforce your growth mindset.

Real-life Examples and Questions

Example 1: A student with a fixed mindset may believe they are "bad at math" and avoid challenging problems. A student with a growth mindset, however, would recognize that their math skills can improve with practice and would seek out opportunities to develop those skills.

Question: Can you recall a time when you held a fixed mindset about a skill or ability? How did it affect your performance, and what steps can you take to adopt a growth mindset in that area?

Example 2: An entrepreneur with a growth mindset views setbacks as learning opportunities and uses those experiences to refine their business strategies. In contrast, an entrepreneur with a fixed mindset may become discouraged and give up after encountering obstacles.

Question: How can you apply the principles of a growth mindset to your personal or professional life? What challenges can you embrace to promote growth and learning?

Suggested Readings and References

Dweck, C. S. (2006). Mindset: The New Psychology of Success. New York: Random House.

Yeager, D. S., & Dweck, C. S. (2012). Mindsets that promote resilience: When students believe that personal characteristics can be developed. Educational Psychologist, 47(4), 302-314.

By cultivating a growth mindset, you will unlock your true potential and set yourself up for success in every aspect of your life. As you continue to develop this mindset, remember that change takes time and consistent effort. Stay committed to growth, embrace challenges, and learn from your experiences. With determination and perseverance, you will transform your mindset and achieve your goals.

In the following sections of this book, we will explore additional strategies and techniques to further develop your success mindset. As an experienced life coach and motivational speaker, I have witnessed firsthand the incredible impact that a growth mindset can have on an individual's personal and professional life. I am confident that by adopting these principles, you too can achieve extraordinary success.

Remember, the journey towards success is an ongoing process, and embracing a growth mindset is the first step towards unlocking your true potential. Stay focused, remain resilient, and never stop learning. Your future self will thank you for the effort you invest in your personal growth today.

So, continue to challenge yourself, seek out new experiences, and embrace the power of a growth mindset. With unwavering determination and a commitment to growth, there's no limit to what you can achieve.

As an experienced consultant and online course creator, I encourage you to further explore the concept of a growth mindset through additional resources, workshops, and coaching. Together, we can cultivate a mindset of success and unleash the extraordinary potential that lies within each of us.

Sarah, a young professional, had always been afraid of failure. She was hesitant to take on new challenges at work, fearing that she might not succeed. As a result, she was often overlooked for promotions and opportunities. After attending a motivational seminar led by an experienced life coach, Sarah learned about the concept of a growth mindset.

Realizing that her fear of failure was holding her back, Sarah decided to embrace the growth mindset and see challenges as opportunities to learn and grow. She started by setting small goals for herself and gradually took on more significant tasks. As she faced obstacles, Sarah reminded herself that setbacks were a natural part of the learning process. Over time, her confidence grew, and she began to excel in her career.

Example 2: Embracing Change

John had been working in the same industry for over 20 years. When his company started implementing new technology, he found it challenging to adapt. He felt overwhelmed by the changes and worried about his job security. An experienced consultant introduced John to the concept of a growth mindset and encouraged him to see change as an opportunity for growth.

John decided to adopt a growth mindset, embracing the new technology and seeking out training to improve his skills. As he became more proficient, he found new ways to contribute to his company's success. By cultivating a growth mindset, John was able to adapt to change and thrive in his career.

Example 3: Overcoming Self-Doubt

Melissa had always wanted to start her own business, but self-doubt held her back. She was afraid that she didn't have what it took to succeed as an entrepreneur. After hearing a motivational speaker discuss the power of a growth mindset, Melissa decided to challenge her limiting beliefs.

She began by setting SMART goals for her business, focusing on developing her skills and knowledge in areas where she felt uncertain. As she faced setbacks, Melissa reminded herself that learning and growth were ongoing processes. With determination and a growth mindset, she was able to overcome her self-doubt and build a successful business.

By embracing a growth mindset and seeing challenges as opportunities for growth, individuals like Sarah, John, and

Melissa were able to overcome their fears, adapt to change, and achieve their goals. With a commitment to growth and a willingness to learn from setbacks, anyone can cultivate a success mindset and unlock their true potential.

Q: What is the difference between a fixed mindset and a growth mindset?

A: A fixed mindset is the belief that intelligence and abilities are static and cannot be improved, while a growth mindset is the belief that one can develop and grow through effort, learning, and perseverance.

Q: How can I cultivate a growth mindset?

A: Cultivating a growth mindset involves embracing challenges, persisting through obstacles, understanding that effort leads to growth, learning from criticism, and being inspired by others' success.

Q: How can a growth mindset improve my performance at work?

A: A growth mindset encourages continuous learning, adaptability, and resilience, which can lead to better problem-solving, innovation, and overall work performance.

Q: Can anyone develop a growth mindset?

A: Yes, anyone can develop a growth mindset through consistent practice, self-awareness, and a commitment to personal growth.

Q: How can I encourage a growth mindset in others?

A: Encourage a growth mindset in others by praising effort, focusing on learning rather than just results, and promoting a culture of continuous improvement and self-development.

Q: How can I change my self-talk to support a growth mindset?

A: Replace negative self-talk with positive affirmations and focus on learning, growth, and effort rather than solely on outcomes.

Q: How can I overcome setbacks with a growth mindset?

A: View setbacks as opportunities to learn and grow, and use them as motivation to work harder and improve.

Q: How can a growth mindset help me deal with criticism?

A: With a growth mindset, you can view criticism as valuable feedback that can help you learn, grow, and improve.

Q: Can adopting a growth mindset improve my relationships?

A: Yes, a growth mindset encourages empathy, understanding, and open communication, which can lead to stronger, healthier relationships.

Q: How can I maintain a growth mindset in the face of adversity?

A: Stay focused on your goals, remind yourself of your past successes, and use adversity as an opportunity to learn and grow.

Q: How does a growth mindset impact motivation?

A: A growth mindset fosters intrinsic motivation, as it emphasizes the importance of effort, learning, and personal development.

Q: How can I set goals that support a growth mindset?

A: Set SMART goals that focus on personal growth, skill development, and learning from experiences.

Q: Can a growth mindset help me overcome fear of failure?

A: Yes, a growth mindset helps you see failure as an opportunity to learn and grow, reducing the fear associated with it.

Q: How can I instill a growth mindset in my children?

A: Encourage effort, curiosity, and learning, praise perseverance, and teach them the value of embracing challenges.

Q: How does a growth mindset contribute to success?

A: A growth mindset fosters resilience, adaptability, and a continuous drive for self-improvement, which are key factors in achieving success.

Q: How long does it take to develop a growth mindset?

A: Developing a growth mindset is an ongoing process that requires commitment, self-awareness, and practice.

Q: How can I identify areas where I have a fixed mindset?

A: Reflect on your beliefs, attitudes, and behaviors, paying attention to areas where you may be resistant to change, struggle with criticism, or fear failure.

Q: How can a growth mindset help me overcome self-doubt?

A: A growth mindset encourages self-confidence through continuous learning and the belief that one can improve and develop over time.

Q: Can a growth mindset help me handle stress better?

A: Yes, a growth mindset can help you view stress as an opportunity for growth and learning, leading to better coping strategies and resilience.

Q: How can I apply a growth mindset to my daily life?

A: Apply a growth mindset by embracing challenges, learning from mistakes, seeking out new experiences, and maintaining a positive outlook on personal growth and development.

Q: What are the key differences between a fixed mindset and a growth mindset?

A: A fixed mindset believes intelligence and talent are predetermined and unchangeable, while a growth mindset views them as developable through effort and learning.

Q: How can I shift from a fixed mindset to a growth mindset?

A: Recognize your fixed mindset tendencies, challenge negative thoughts, embrace challenges, learn from mistakes, and seek feedback to help you grow.

Q: Can a growth mindset improve my relationships?

A: Yes, a growth mindset can lead to better communication, understanding, and empathy, as it encourages open-mindedness and adaptability in relationships.

Q: How does having a growth mindset contribute to success?

A: A growth mindset fosters resilience, adaptability, and a love for learning, which are essential for overcoming obstacles and achieving success.

Q: How can I help my children develop a growth mindset?

A: Encourage effort and perseverance, praise progress, emphasize learning over grades, and model a growth mindset through your actions and words.

Q: How can I maintain a growth mindset at work?

A: Embrace challenges, seek feedback, view setbacks as opportunities to learn, and focus on continuous improvement in your skills and abilities.

Q: Can adopting a growth mindset help me overcome fear of failure?

A: Yes, a growth mindset reframes failure as an opportunity for learning and growth, allowing you to face challenges without fear.

Q: What role does self-reflection play in cultivating a growth mindset?

A: Self-reflection helps identify fixed mindset tendencies, recognize personal growth opportunities, and develop strategies to cultivate a growth mindset.

Q: Can a growth mindset help me achieve my goals?

A: Absolutely, a growth mindset encourages perseverance, learning, and adaptability, which are critical for achieving goals.

Q: How do I know if I'm making progress in developing a growth mindset?

A: Signs of progress include embracing challenges, learning from mistakes, seeking feedback, and exhibiting resilience in the face of setbacks.

Q: Can a growth mindset improve my mental health?

A: Yes, adopting a growth mindset can reduce stress, enhance resilience, and promote a more optimistic outlook on life.

Q: How can I develop a growth mindset in my team or organization?

A: Encourage a culture of learning, support risk-taking, provide constructive feedback, and celebrate effort and progress rather than only results.

Q: How long does it take to cultivate a growth mindset?

A: Developing a growth mindset is an ongoing process that requires consistent effort, self-awareness, and reflection.

Q: Are there any books or resources to help me learn more about growth mindset?

A: Yes, "Mindset: The New Psychology of Success" by Carol Dweck is a popular resource, along with various articles, TED talks, and podcasts on the subject.

Q: Can a growth mindset help with procrastination?

A: Yes, by viewing tasks as opportunities for growth and learning, a growth mindset can help you overcome procrastination.

Q: How does a growth mindset impact creativity and innovation?

A: A growth mindset encourages risk-taking, experimentation, and learning from failures, all of which are essential for creativity and innovation.

Q: Can a growth mindset make me more receptive to feedback?

A: Yes, by viewing feedback as an opportunity for growth, a growth mindset helps you become more open to receiving and acting on constructive criticism.

Q: How can I incorporate a growth mindset into my daily routine?

A: Reflect on your thoughts and actions, challenge negative beliefs

Q: What are the key differences between a fixed mindset and a growth mindset?

A: A fixed mindset believes intelligence and talent are predetermined and unchangeable, while a growth mindset views them as developable through effort and learning.

Q: How can I shift from a fixed mindset to a growth mindset?

A: Recognize your fixed mindset tendencies, challenge negative thoughts, embrace challenges, learn from mistakes, and seek feedback to help you grow.

Q: Can a growth mindset improve my relationships?

A: Yes, a growth mindset can lead to better communication, understanding, and empathy, as it encourages open-mindedness and adaptability in relationships.

Q: How does having a growth mindset contribute to success?

A: A growth mindset fosters resilience, adaptability, and a love for learning, which are essential for overcoming obstacles and achieving success.

Q: How can I help my children develop a growth mindset?

A: Encourage effort and perseverance, praise progress, emphasize learning over grades, and model a growth mindset through your actions and words.

Q: How can I maintain a growth mindset at work?

A: Embrace challenges, seek feedback, view setbacks as opportunities to learn, and focus on continuous improvement in your skills and abilities.

Q: Can adopting a growth mindset help me overcome fear of failure?

A: Yes, a growth mindset reframes failure as an opportunity for learning and growth, allowing you to face challenges without fear.

Q: What role does self-reflection play in cultivating a growth mindset?

A: Self-reflection helps identify fixed mindset tendencies, recognize personal growth opportunities, and develop strategies to cultivate a growth mindset.

Q: Can a growth mindset help me achieve my goals?

A: Absolutely, a growth mindset encourages perseverance, learning, and adaptability, which are critical for achieving goals.

Q: How do I know if I'm making progress in developing a growth mindset?

A: Signs of progress include embracing challenges, learning from mistakes, seeking feedback, and exhibiting resilience in the face of setbacks.

Q: Can a growth mindset improve my mental health?

A: Yes, adopting a growth mindset can reduce stress, enhance resilience, and promote a more optimistic outlook on life.

Q: How can I develop a growth mindset in my team or organization?

A: Encourage a culture of learning, support risk-taking, provide constructive feedback, and celebrate effort and progress rather than only results.

Q: How long does it take to cultivate a growth mindset?

A: Developing a growth mindset is an ongoing process that requires consistent effort, self-awareness, and reflection.

Q: Are there any books or resources to help me learn more about growth mindset?

A: Yes, "Mindset: The New Psychology of Success" by Carol Dweck is a popular resource, along with various articles, TED talks, and podcasts on the subject.

Q: Can a growth mindset help with procrastination?

A: Yes, by viewing tasks as opportunities for growth and learning, a growth mindset can help you overcome procrastination.

Q: How does a growth mindset impact creativity and innovation?

A: A growth mindset encourages risk-taking, experimentation, and learning from failures, all of which are essential for creativity and innovation.

Q: Can a growth mindset make me more receptive to feedback?

A: Yes, by viewing feedback as an opportunity for growth, a growth mindset helps you become more open to receiving and acting on constructive criticism.

Q: How can I incorporate a growth mindset into my daily routine?

A: Reflect on your thoughts and actions, challenge negative beliefs and assumptions, seek learning opportunities, embrace challenges, and practice gratitude for personal growth experiences.

Q: Can a growth mindset help me in dealing with stress and adversity?

A: Yes, a growth mindset can help you see stress and adversity as opportunities for growth, allowing you to develop resilience and better cope with challenges.

Q: How can I encourage a growth mindset in others around me?

A: Share your own growth mindset experiences, praise effort and progress, offer constructive feedback, and provide support for learning and personal development.

Techniques and Tools for Cultivating a Growth Mindset

As a Life Coach, I've found that these techniques and tools can be highly effective in cultivating a growth mindset. Follow these steps and incorporate them into your daily routine to help you develop a success mindset.

Reflect on Your Mindset: Take a moment to assess your current mindset. Do you see challenges as opportunities or threats? Understanding your mindset is the first step in making a change.

Mindful Awareness: Practice mindfulness to become more aware of your thoughts and emotions. This can help you identify limiting beliefs and replace them with more empowering ones.

Affirmations: Use positive affirmations to reinforce a growth mindset. Repeat phrases such as "I can learn from my mistakes" or "I embrace challenges" daily to internalize these beliefs.

Journaling: Write down your thoughts, feelings, and experiences related to growth and learning. This will help you recognize patterns and track your progress over time.

Embrace Failure: View failures as learning opportunities instead of setbacks. Analyze what went wrong and use that knowledge to improve in the future.

Seek Feedback: Actively seek feedback from others and view it as a valuable resource for improvement. Be open to constructive criticism and apply it to your personal growth.

Set Learning Goals: Instead of focusing solely on performance goals, set learning goals that prioritize the development of new skills or knowledge.

Challenge Yourself: Step outside your comfort zone and tackle new challenges that will promote growth and learning.

Surround Yourself with Growth-Minded People: Build a support network of individuals who share a growth mindset and can provide encouragement and inspiration.

Celebrate Progress: Acknowledge and celebrate your progress, no matter how small. This reinforces the importance of personal growth and keeps you motivated.

By following these steps and incorporating these tools into your daily life, you'll be well on your way to cultivating a growth mindset and achieving lasting success. Remember, personal growth is a continuous journey, so stay committed to the process and enjoy the journey.

EMBRACING THE POWER OF POSITIVE THINKING

The power of positive thinking cannot be underestimated when it comes to developing a success mindset. As an experienced Life Coach and Motivational Speaker, I have seen firsthand the impact that positive thinking can have on a person's ability to overcome obstacles, achieve their goals, and maintain a strong sense of self-worth. In this section, we will explore the importance of positive thinking, discuss strategies for cultivating a positive mindset, and provide real-life examples of individuals who have harnessed the power of positivity to achieve remarkable results.

Positive thinking is about more than just blind optimism; it's about embracing a mindset that focuses on opportunities, growth, and potential rather than dwelling on negativity, limitations, and past failures. This way of thinking has been proven to improve mental and physical health, increase resilience, and promote overall well-being.

To cultivate a positive mindset, consider incorporating the following strategies into your daily routine:

Practice gratitude: Each day, make a list of at least three things you are grateful for. This simple exercise can help shift your focus away from negativity and promote an appreciation for the good things in your life.

Reframe negative thoughts: When faced with a negative thought, challenge it by looking for alternative, more positive interpretations of the situation. For example, if you catch yourself thinking, "I'll never be able to do this," try reframing it as, "This may be challenging, but I can learn and grow from the experience."

Surround yourself with positivity: Spend time with positive people and engage in activities that bring you joy and fulfillment. Avoid negative influences, such as toxic relationships or excessive consumption of negative news.

Set realistic goals: Break down your larger goals into smaller, achievable steps. This approach can help you maintain a positive outlook by allowing you to experience success and progress along the way.

Visualize success: Spend time each day visualizing yourself achieving your goals. Imagine the feelings of pride, satisfaction, and happiness that will accompany your accomplishments.

One real-life example of the power of positive thinking can be found in the story of John, a middle-aged man who lost his job during a company downsizing. Rather than dwelling on the loss and letting negativity consume him, John chose to view the situation as an opportunity for personal growth and reinvention. He invested time in learning new skills, networking, and eventually found a more fulfilling job in a different industry. By maintaining a positive mindset, John was able to turn a challenging situation into a transformative experience.

Another example comes from Sarah, a young woman who struggled with low self-esteem and body image issues. Through therapy and the support of her Life Coach, Sarah began to embrace positive thinking and learned to appreciate her body and its capabilities. She started participating in physical activities that brought her joy and eventually became a fitness instructor, inspiring others to lead healthy, confident lives.

In conclusion, embracing the power of positive thinking can be a transformative force in our lives. By cultivating a positive mindset and implementing the strategies discussed in this section, you can unlock your full potential and achieve lasting success. Remember to always seek growth, focus on the good, and never underestimate the power of a positive attitude. For further reading on this topic, I recommend the following books: "The Power of Positive Thinking" by Norman Vincent Peale and "Learned Optimism" by Martin E.P. Seligman.

Example 1: Overcoming Career Stagnation with Positive Thinking

Michelle, a talented graphic designer, felt stuck in her career for years. She believed that she couldn't achieve more success due to her age and the competitive nature of her industry. This negative mindset prevented her from seeking new opportunities, and her dissatisfaction grew.

Michelle decided to consult a Life Coach, who helped her shift her mindset and embrace the power of positive thinking. She began to focus on her strengths and unique experiences instead of dwelling on her perceived limitations. By doing so, Michelle gained the confidence to apply for a higher position at a prestigious design firm. She landed the job and now enjoys a more fulfilling and challenging career.

Problem: Career stagnation due to negative mindset

Solution: Embracing positive thinking and focusing on strengths

Example 2: Overcoming Self-Doubt in a New Venture

David had always dreamed of starting his own business but was paralyzed by self-doubt and fear of failure. He worried that he lacked the necessary skills and resources to make his dream a reality. This negative mindset held him back from pursuing his passion.

After attending a motivational seminar by an Experienced Consultant, David decided to change his mindset and embrace positive thinking. He focused on the possibilities and potential of his business idea, allowing himself to envision a successful outcome. David also sought guidance from mentors and invested in professional development to build his confidence and skills.

With a newfound belief in himself and his abilities, David launched his business, and it quickly became a success. He attributes this achievement to the power of positive thinking and the support he received from his mentors.

Problem: Fear of failure and self-doubt in starting a business

Solution: Embracing positive thinking and seeking mentorship

Example 3: Overcoming Procrastination and Achieving Personal Goals

Emma, a young professional, struggled with procrastination, which prevented her from achieving her personal goals. She often felt overwhelmed and believed she could never accomplish everything she set out to do, leading to further procrastination.

During an online course created by an experienced Life Coach, Emma learned about the power of positive thinking and how it could help her overcome her procrastination. She began to practice reframing her negative thoughts and focusing on the progress she could make each day, no matter how small.

By embracing a positive mindset, Emma gradually overcame her procrastination and began to make significant progress towards her goals. She now feels more in control of her life and is excited about the future.

Problem: Procrastination and feeling overwhelmed by personal goals

Solution: Embracing positive thinking and focusing on daily progress

Example 4: Conquering Public Speaking Anxiety

John, an accomplished entrepreneur, struggled with severe anxiety when speaking in public. This fear limited his ability to share his expertise and connect with potential clients, holding back his business's growth.

Determined to conquer his anxiety, John sought the help of an experienced motivational speaker, who introduced him to the power of positive thinking. They worked together to change John's mindset, focusing on his strengths as a speaker and envisioning successful presentations.

As John practiced positive thinking techniques, his anxiety began to subside. He started taking on more speaking engagements and found that the more he spoke, the more confident he became. Today, John is a sought-after speaker in his industry and credits positive thinking for his transformation.

Problem: Public speaking anxiety

Solution: Embracing positive thinking and focusing on strengths

Example 5: Overcoming Relationship Challenges with Positive Thinking

Laura and James, a married couple, were struggling with constant arguing and miscommunication. Their negative mindset toward each other was driving them apart, and they feared their relationship might not survive.

Upon the recommendation of a close friend, they decided to attend a couples' workshop led by an experienced Life Coach. The workshop focused on teaching the power of positive thinking and communication techniques to improve their relationship.

Through the workshop, Laura and James learned to replace their negative thoughts with positive ones, which significantly improved their communication and understanding. They began to focus on the love they shared and the desire to grow together. As a result, their relationship flourished, and they now enjoy a stronger bond than ever before.

Problem: Relationship challenges due to negative thinking and miscommunication

Solution: Embracing positive thinking and improving communication techniques

By sharing these real-life examples, readers can see how embracing the power of positive thinking can lead to personal and professional growth. These stories demonstrate the potential for change when individuals commit to adopting a more optimistic mindset and seeking guidance from experienced professionals.

Technique 1: Positive Affirmations

Positive affirmations are powerful statements that help reprogram your mindset and build self-confidence. Start by creating a list of affirmations tailored to your goals and aspirations. Repeat these affirmations daily, either out loud or in your mind. Examples of positive affirmations include:

I am capable of achieving my goals.

I am worthy of success and happiness.

I am a positive influence on those around me.

Technique 2: Gratitude Journaling

Gratitude journaling is a powerful way to shift your focus from negative thoughts to positive ones. Each day, write down at least three things you are grateful for. This practice will help you develop a more optimistic outlook on life.

Technique 3: Visualization

Visualization is a powerful tool that can help you embrace positive thinking. Close your eyes and imagine yourself achieving your goals or overcoming challenges. By visualizing success, you train your mind to expect positive outcomes, increasing your confidence and motivation.

Technique 4: Mindfulness Meditation

Mindfulness meditation helps you become more aware of your thoughts and feelings, enabling you to identify and replace negative thoughts with positive ones. Set aside 10-15 minutes each day to practice mindfulness meditation. Focus on your breath, and gently bring your attention back to the present moment whenever your mind wanders.

Technique 5: Surround Yourself with Positivity

Create a positive environment by surrounding yourself with uplifting people, books, podcasts, and other resources. This will help you reinforce positive thinking and maintain an optimistic outlook.

Technique 6: Challenge Negative Thoughts

When negative thoughts arise, challenge them by asking yourself if they are based on facts or irrational fears. Replace these thoughts with more balanced and positive ones.

Technique 7: Set SMART Goals

Setting Specific, Measurable, Achievable, Relevant, and Time-bound (SMART) goals will help you maintain a positive mindset by giving you a clear path to success. Break your goals down into smaller steps and celebrate your progress along the way.

These techniques, when practiced consistently, can help you embrace the power of positive thinking and develop a success mindset. By following these steps, you'll be well on your way to achieving your goals and living a more fulfilling life.

Q: What is the main benefit of embracing positive thinking?

A: Positive thinking can improve your mental well-being, increase your motivation, and enhance your resilience, ultimately leading to a more successful and fulfilling life.

Q: How can I create my own positive affirmations?

A: Reflect on your goals, values, and strengths, and craft statements that highlight these aspects in a positive, empowering way.

Q: How does gratitude journaling contribute to a positive mindset?

A: Gratitude journaling helps shift your focus from negative thoughts to positive ones, fostering a more optimistic outlook on life.

Q: Can visualization techniques really impact my success?

A: Yes, visualization helps train your mind to expect positive outcomes, which can increase your confidence, motivation, and likelihood of success.

Q: How can I practice mindfulness meditation?

A: Set aside 10-15 minutes each day to focus on your breath and stay present, gently bringing your attention back to the moment whenever your mind wanders.

Q: Why is it essential to surround myself with positivity?

A: A positive environment reinforces optimistic thinking and helps maintain an overall positive mindset.

Q: How can I challenge negative thoughts effectively?

A: Analyze whether your negative thoughts are based on facts or irrational fears, and replace them with more balanced and positive ones.

Q: What are SMART goals, and how can they help me develop a success mindset?

A: SMART goals are Specific, Measurable, Achievable, Relevant, and Time-bound, providing a clear path to success and promoting a positive mindset.

Q: How can I apply the power of positive thinking to my professional life?

A: Focus on your strengths, set SMART goals, visualize your success, and surround yourself with positive influences to maintain an optimistic attitude in your professional endeavors.

Q: Can positive thinking help me overcome failures and setbacks?

A: Yes, a positive mindset enables you to view challenges as opportunities for growth and learning, making it easier to bounce back from setbacks.

Q: How can I maintain a positive mindset during difficult times?

A: Practice gratitude, focus on your goals, use positive affirmations, and seek support from uplifting people to help you navigate through challenging situations.

Q: Can positive thinking improve my relationships with others?

A: Absolutely, a positive mindset can help you communicate more effectively, empathize better, and foster stronger connections with others.

Q: How can I encourage positive thinking in my team or family members?

A: Share your own positive experiences, create a supportive environment, and provide resources that promote positivity and growth.

Q: Is it possible to be overly positive?

A: Yes, excessive positivity can lead to unrealistic expectations and denial of real problems. Balance is key; acknowledge challenges while maintaining an optimistic outlook.

Q: How long does it take to see the benefits of positive thinking?

A: The benefits of positive thinking can be experienced immediately, but long-term changes require consistent practice and effort.

Q: How can I measure my progress in developing a positive mindset?

A: Reflect on your thought patterns, reactions to challenges, and overall well-being to gauge improvements in your mindset.

Q: Can positive thinking help reduce stress and anxiety?

A: Yes, cultivating a positive mindset can help you manage stress and anxiety more effectively by promoting resilience and a balanced perspective.

Q: What role does self-compassion play in positive thinking?

A: Self-compassion encourages a kind and understanding attitude towards yourself, which can help you replace negative thoughts with more positive ones.

Q: How can I create a daily routine that supports a positive mindset?

A: Incorporate practices such as gratitude journaling, meditation, affirmations, and exercise into your daily routine to reinforce positive thinking and overall well-being.

Q: What is the connection between physical health and positive thinking?

A: A healthy lifestyle, including proper nutrition, exercise, and sleep, can contribute to a positive mindset by improving mood, increasing energy levels, and reducing stress.

Mastering Self-Discipline and Time Management

As a life coach, I've seen time and time again that the cornerstone of a successful mindset is a combination of self-discipline and effective time management. The ability to prioritize tasks, set goals, and maintain focus on what truly matters will greatly increase your chances of achieving your dreams. In this section, we will dive into the essential principles of self-discipline and time management to help you take control of your life.

To begin, let's define self-discipline. Self-discipline is the ability to control one's impulses, emotions, and behaviors to achieve long-term goals. It involves making choices that align with your values and aspirations, even when faced with distractions or challenges. Time management, on the other hand, is the process of organizing and planning how to allocate your time effectively to maximize productivity and achieve your goals.

The following strategies will help you improve your self-discipline and time management skills:

Set clear, realistic goals: Begin by identifying your long-term goals and breaking them down into smaller, more manageable objectives. This will give you a clear roadmap to follow and help you maintain focus on what truly matters.

Prioritize tasks: Make a list of tasks you need to complete and rank them according to their importance and urgency. This will help you allocate your time and energy to tasks that align with your goals.

Establish a routine: Create a daily routine that includes time for work, leisure, exercise, and self-improvement activities. This will help you maintain balance in your life and ensure that you dedicate sufficient time to achieving your goals.

Eliminate distractions: Identify and remove any distractions that could hinder your progress. This may include turning off notifications on your devices, setting boundaries with friends and family, or decluttering your workspace.

Practice self-awareness: Regularly evaluate your progress and identify areas where you can improve. By recognizing your weaknesses and strengths, you can make adjustments to your approach and maintain a growth mindset.

Some suggested readings to further enhance your understanding of self-discipline and time management include:

"The Miracle Morning" by Hal Elrod

"The One Thing" by Gary Keller and Jay Papasan

"Eat That Frog!" by Brian Tracy

Real-life example:

Samantha, a budding entrepreneur, was struggling to find enough hours in the day to manage her business, maintain a healthy lifestyle, and spend quality time with her family. With the help of a life coach, Samantha was able to set clear goals, prioritize tasks, and create a daily routine that allowed her to balance her work and personal life. Through improved self-discipline and time management, she was able to grow her business while still dedicating time to her family and personal well-being.

Question: How can I become more self-disciplined in my daily life?

Answer: Start by setting clear, realistic goals and breaking them down into smaller, manageable tasks. Develop a daily routine that incorporates time for work, self-improvement, and leisure. Practice self-awareness to identify areas where you can improve and eliminate distractions that hinder your progress.

Example 1: Overcoming Procrastination

Problem: Tom, a talented graphic designer, struggled with procrastination. He found himself constantly putting off important tasks and missing deadlines, which negatively affected his career and personal life.

Question: How can Tom overcome his procrastination and improve his self-discipline?

Answer: As a life coach, I would advise Tom to identify the root causes of his procrastination, such as fear of failure or lack of motivation. To improve his self-discipline, he could set clear goals, break them down into smaller tasks, and create a daily routine. Additionally, Tom could implement the Pomodoro Technique, which involves working in focused, timed intervals with short breaks in between to improve productivity and combat procrastination.

Example 2: Balancing Work and Personal Life

Problem: Sarah, a busy working mother, found it challenging to balance her demanding job and personal life. She felt overwhelmed and unable to dedicate enough time to her family, friends, and self-care.

Question: How can Sarah improve her time management skills to create a better work-life balance?

Answer: As a motivational speaker, I would suggest that Sarah start by setting boundaries with her work, such as limiting after-hours emails and calls. She should prioritize tasks based on their importance and create a schedule that includes time for her family, friends, and self-care. By developing a routine and sticking to it, Sarah can improve her time management skills and achieve a more balanced life.

Example 3: Staying Consistent with Fitness Goals

Problem: Jake, a busy professional, wanted to improve his physical fitness but struggled to maintain consistency in his exercise routine due to his hectic schedule.

Question: How can Jake develop better self-discipline and time management skills to achieve his fitness goals?

Answer: As an experienced consultant, I would recommend that Jake set specific, measurable, and realistic fitness goals to stay motivated. To improve his self-discipline, he could schedule workouts during times when he's most likely to follow through, such as early mornings or lunch breaks. By incorporating exercise into his daily routine and tracking his progress, Jake can stay accountable and committed to his fitness journey.

Technique 1: The Prioritization Matrix

To master self-discipline and time management, start by prioritizing tasks using a prioritization matrix. This matrix helps you categorize tasks based on their urgency and importance, allowing you to focus on what matters most.

Step 1: Create a 2x2 matrix with labels "Urgent" and "Not Urgent" on the horizontal axis, and "Important" and "Not Important" on the vertical axis.

Step 2: List your tasks and assign them to the appropriate quadrant.

Step 3: Focus on completing tasks in the "Urgent and Important" quadrant first, then move on to the "Not Urgent but Important" tasks.

Step 4: Re-evaluate and update the matrix regularly to ensure you're always working on high-priority tasks.

Technique 2: The Time Blocking Method

Time blocking is a technique that involves scheduling specific blocks of time for each task throughout the day. This method helps improve focus and productivity.

Step 1: Create a daily or weekly schedule template, dividing your day into 30-minute or 1-hour blocks.

Step 2: Assign tasks to specific time blocks, ensuring you allocate enough time for each task.

Step 3: Include breaks and buffer time for unexpected tasks or interruptions.

Step 4: Commit to working on the assigned task during its designated time block, minimizing distractions.

Technique 3: The 2-Minute Rule

The 2-Minute Rule states that if a task takes less than two minutes to complete, do it immediately. This technique helps you quickly tackle small tasks, preventing them from accumulating and overwhelming you.

Step 1: Identify tasks that can be completed in two minutes or less.

Step 2: Complete the task immediately, without procrastination.

Step 3: If a task takes longer than two minutes, either schedule it for later or break it down into smaller, two-minute tasks.

Technique 4: The Accountability Partner

An accountability partner is someone who helps you stay committed to your goals by holding you responsible for your actions. This partnership can improve self-discipline and time management by providing external motivation.

Step 1: Find a reliable and supportive accountability partner, such as a friend, family member, or colleague.

Step 2: Share your goals, plans, and deadlines with your partner.

Step 3: Regularly check in with your partner to discuss progress, challenges, and successes.

Step 4: Offer support and encouragement to your partner, and seek their guidance when needed.

These techniques and tools, when used consistently, can help you develop a success mindset by mastering self-discipline and time management. Practice them regularly and adjust as needed to find the strategies that work best for you.

Q: How can I improve my self-discipline to better manage my time?

A: Start by setting clear goals, breaking them down into smaller tasks, and creating a daily or weekly schedule. Incorporate techniques such as time blocking, the prioritization matrix, and the 2-minute rule to help you stay focused and disciplined.

Q: What is the prioritization matrix, and how can it help me?

A: The prioritization matrix is a tool that categorizes tasks based on their urgency and importance. It helps you focus on high-priority tasks first, ultimately improving your time management.

Q: How can I stay focused while working on a task?

A: Minimize distractions, create a conducive work environment, and use techniques like time blocking to allocate specific time periods for each task.

Q: What is time blocking, and how does it improve time management?

A: Time blocking is a technique that involves scheduling specific blocks of time for each task throughout the day. It helps improve focus and productivity by ensuring you dedicate the necessary time to each task.

Q: How can I avoid procrastination?

A: Break tasks into smaller steps, set realistic deadlines, and use the 2-minute rule for quick tasks. An accountability partner can also provide external motivation to help you stay on track.

Q: What role does self-discipline play in time management?

A: Self-discipline helps you stay focused on your goals, resist distractions, and maintain consistency in your efforts, ultimately improving your time management skills.

Q: How can an accountability partner help me manage my time better?

A: An accountability partner provides external motivation, encouragement, and support, helping you stay committed to your goals and maintain self-discipline.

Q: Can meditation or mindfulness practices improve my time management skills?

A: Yes, meditation and mindfulness can improve focus, reduce stress, and enhance self-discipline, all of which contribute to better time management.

Q: How can I balance my work and personal life while maintaining self-discipline?

A: Create a clear boundary between work and personal time, prioritize self-care, and use effective time management techniques to ensure you allocate time for both aspects of your life.

Q: Can setting goals help me improve my self-discipline and time management?

A: Yes, setting specific, measurable, achievable, relevant, and time-bound (SMART) goals provides direction and motivation, enhancing self-discipline and time management.

Q: How can I stay motivated to maintain self-discipline and manage my time effectively?

A: Focus on your long-term goals, celebrate small successes, and surround yourself with positive influences to maintain motivation.

Q: What is the 2-minute rule, and how does it help with time management?

A: The 2-minute rule states that if a task takes less than two minutes to complete, do it immediately. This technique helps you quickly tackle small tasks, preventing them from accumulating and overwhelming you.

Q: How can I manage unexpected tasks or interruptions without compromising my self-discipline and time management?

A: Include buffer time in your schedule for unexpected tasks, and learn to prioritize or delegate tasks when necessary.

Q: What are some ways to track and measure my progress in self-discipline and time management?

A: Use a planner, journal, or digital tool to track your tasks, deadlines, and accomplishments. Regularly evaluate your progress and adjust your strategies as needed.

Q: Can multitasking improve my time management skills?

A: Multitasking often reduces focus and productivity. Instead, concentrate on one task at a time using techniques like time blocking for better time management.

Q: How can I manage my energy levels to maintain self-discipline and time management?

A: Pay attention to your body's natural rhythms and schedule tasks during your most productive times of day. Also, prioritize self-care, exercise, and proper nutrition to maintain your energy levels.

Q: How can I overcome feelings of overwhelm while trying to maintain self-discipline and manage my time?

A: Break tasks into smaller, manageable steps, prioritize tasks, and focus on one task at a time. Remember to take breaks and practice self-compassion when feeling overwhelmed.

Q: What role do habits play in self-discipline and time management?

A: Developing good habits, such as consistent morning and evening routines, can help you maintain self-discipline and manage your time effectively by automating certain aspects of your life.

Q: How can I identify and eliminate time-wasting activities to improve my self-discipline and time management?

A: Track your daily activities to identify patterns of time-wasting behavior. Once identified, minimize or eliminate these activities by setting boundaries or using techniques like the Pomodoro Technique to maintain focus.

Q: Can delegating tasks improve my self-discipline and time management skills?

A: Yes, delegating tasks can help you focus on high-priority tasks that require your expertise, allowing you to better manage your time and maintain self-discipline.

Leveraging Visualization Techniques

As an experienced life coach, I've seen the power of visualization techniques in helping clients achieve their goals and develop a success mindset. Visualization is a powerful mental tool that can help you bring your dreams to life by creating vivid mental images of your desired outcome. In this section, we'll discuss the benefits of visualization and provide practical tips on incorporating these techniques into your daily routine.

One of the main benefits of visualization is that it allows you to create a mental blueprint of your goals. When you can clearly see the outcome you desire, your brain works to find ways to bridge the gap between your current reality and your envisioned future. Visualization also helps boost motivation, increases self-confidence, and enhances focus, all of which are critical factors for developing a success mindset.

To effectively use visualization techniques, follow these steps:

Set clear, specific goals: To visualize effectively, you need to have a clear idea of what you want to achieve. Write down your goals, making sure they are specific, measurable, and realistic.

Create a quiet space: Find a comfortable place where you can be free from distractions, allowing you to fully focus on your visualization practice.

Close your eyes and relax: Take a few deep breaths, and allow your mind and body to become calm and relaxed.

Use all your senses: As you visualize your desired outcome, engage all your senses. Imagine what you see, hear, smell, taste, and feel as if you have already achieved your goal.

Add emotion: Connect with the emotions you would experience once you've achieved your goals. This emotional connection will help solidify the visualization in your mind and enhance your motivation.

Practice regularly: Visualization is most effective when practiced consistently. Set aside time each day for your visualization practice, even if it's just a few minutes.

To further support your visualization practice, consider creating a vision board with images and words that represent your goals. This can be a physical board or a digital one, using apps or software designed for this purpose. Regularly reviewing your vision board can help reinforce your visualization practice and maintain your motivation.

Real-life example:

Sarah, a client of mine, struggled with self-confidence and was hesitant to pursue her dream of starting her own business. Through visualization techniques, Sarah was able to see herself running a successful business, interacting with clients, and experiencing the freedom and fulfillment she longed for. This mental rehearsal helped her build the confidence needed to take action, and she eventually launched her business, which has been thriving.

Remember, visualization is a powerful tool for developing a success mindset. By regularly practicing visualization techniques, you can create a clear mental image of your goals, boost your motivation, and enhance your focus on the path to achieving your dreams.

Suggested readings:

"Creative Visualization: Use the Power of Your Imagination to Create What You Want in Your Life" by Shakti Gawain

"The Power of Your Subconscious Mind" by Joseph Murphy

"The Miracle Morning: The Not-So-Obvious Secret Guaranteed to Transform Your Life (Before 8AM)" by Hal Elrod

References:

Latham, G. P., & Locke, E. A. (1991). Self-regulation through goal setting. Organizational Behavior and Human Decision Processes, 50(2), 212-247.

Kosslyn, S. M., Ganis, G., & Thompson, W. L. (2001). Neural foundations of imagery. Nature Reviews Neuroscience, 2(9), 635-642.

Real-life story 1: Overcoming Procrastination with Visualization

Jack, a college student, struggled with procrastination and time management, causing him to fall behind in his studies. As his life coach, I recommended he try visualization techniques to improve his motivation and focus. Jack began visualizing himself successfully completing assignments, studying for exams, and confidently participating in class discussions.

After several weeks of consistent visualization practice, Jack noticed a significant improvement in his motivation and ability to concentrate on his studies. By picturing himself as a successful student, he was able to transform his mindset and overcome his procrastination habit.

Real-life story 2: Achieving Weight Loss Goals through Visualization

Emma, a young professional, had been struggling to lose weight for years. She tried various diets and exercise programs but was never able to maintain her motivation. As her motivational speaker, I encouraged Emma to incorporate visualization techniques into her daily routine.

Emma began visualizing herself as fit, healthy, and confident, feeling good in her body. She also imagined herself enjoying healthy meals and participating in regular exercise activities. As a result, Emma's motivation to stick to her diet and exercise plan increased, and she eventually achieved her weight loss goal.

Real-life story 3: Boosting Sales Performance with Visualization

As an experienced consultant, I was approached by a sales team that was struggling to meet their monthly targets. I introduced them to the power of visualization as a tool to improve their performance.

The sales team members started visualizing themselves successfully closing deals, confidently handling objections, and building strong relationships with clients. They also imagined themselves celebrating their successes together. This practice helped them develop a success mindset, and within a few months, they began consistently meeting and exceeding their sales targets.

Real-life story 4: Overcoming Fear of Public Speaking through Visualization

Samantha, a talented professional, was offered a promotion that required her to give regular presentations to large audiences. However, she suffered from a crippling fear of public speaking. As her life coach, I suggested Samantha try visualization techniques to help her overcome this fear.

Samantha began visualizing herself confidently delivering her presentations, engaging her audience, and receiving positive feedback. She practiced this visualization daily, and over time, her fear of public speaking diminished. Eventually, Samantha was able to accept the promotion and excel in her new role.

These real-life stories demonstrate the power of visualization techniques in helping individuals overcome challenges, achieve their goals, and develop a success mindset. By incorporating visualization into your daily routine, you too can experience the transformative power of this practice.

Question 1: How can visualization techniques help me achieve my goals?

Answer: Visualization techniques help you create a mental image of your desired outcome, which in turn boosts your motivation, focus, and self-confidence. By regularly picturing yourself achieving your goals, you train your mind to believe in your success, making it easier to take action and stay committed.

Question 2: How often should I practice visualization techniques?

Answer: For best results, practice visualization daily, ideally in a quiet, relaxed setting. Consistency is key, as the more you practice, the more effective visualization becomes in shaping your mindset and behavior.

Question 3: Can visualization techniques help with anxiety and stress?

Answer: Yes, visualization can help alleviate anxiety and stress by allowing you to imagine yourself in a calm, peaceful state or successfully navigating challenging situations. This mental rehearsal helps build your resilience and coping skills.

Question 4: Are there specific visualization exercises I can use to improve my performance in sports or other activities?

Answer: Absolutely. Many athletes and performers use visualization to mentally rehearse their movements and techniques, which helps improve their actual performance. Visualize yourself executing the task flawlessly and confidently, and your mind will become better attuned to achieving that outcome.

Question 5: How long should each visualization session last?

Answer: Visualization sessions can range from a few minutes to 20 minutes or more, depending on your preference and the complexity of the goal or task. The key is to focus on quality rather than quantity, ensuring that your mental images are clear, vivid, and detailed.

Question 6: Can I use visualization to improve my relationships and communication skills?

Answer: Yes, visualization can be used to enhance your interpersonal skills by mentally rehearsing successful interactions, empathetic listening, and assertive communication. By practicing these scenarios in your mind, you can build confidence and develop more effective communication habits.

Question 7: How can I make my visualization practice more effective?

Answer: To enhance your visualization practice, make your mental images as detailed and vivid as possible, incorporating all your senses. Additionally, try to evoke the emotions associated with achieving your goal, as this helps strengthen the connection between your visualization and your motivation.

Question 8: Is it normal to struggle with maintaining focus during visualization exercises?

Answer: Yes, it's common for people to experience difficulty focusing during visualization, especially when they first start. As you practice more consistently, you'll likely find it easier to maintain focus and create vivid mental images.

Question 9: Can I use visualization techniques to help me break bad habits?

Answer: Absolutely. Visualization can be a powerful tool for replacing unhealthy habits with positive ones. Visualize yourself successfully resisting the urge to engage in the bad habit and instead choosing a healthier alternative. This mental rehearsal can help rewire your brain and support lasting change.

Question 10: What should I do if I find it difficult to create clear mental images during visualization?

Answer: If you struggle with creating mental images, try focusing on the feelings or sensations associated with achieving your goal. You can also try using guided visualization recordings, which can help provide a more structured and detailed framework for your practice.

Question 11: Can visualization help with my decision-making process?

Answer: Yes, visualization can assist you in exploring different options and their potential outcomes. By mentally rehearsing each choice, you can gain insights into how each decision might impact your life and make more informed choices.

Question 12: How can visualization techniques help me become more confident?

Answer: Visualization allows you to mentally rehearse situations in which you display confidence, effectively training your brain to adopt this mindset in real life. As you consistently visualize yourself acting confidently, you'll find it easier to embody that trait in your day-to-day interactions.

Question 13: Can visualization help me improve my public speaking skills?

Answer: Yes, visualization can be a valuable tool in enhancing your public speaking abilities. By mentally rehearsing your presentation, visualizing yourself speaking confidently and engaging your audience, you can reduce anxiety and improve your performance.

Question 14: Can I combine visualization techniques with other personal development strategies?

Answer: Absolutely! Combining visualization with other personal development techniques, such as goal setting, affirmations, and meditation, can create a powerful synergy that amplifies your results and accelerates your progress.

Question 15: How can I stay motivated to consistently practice visualization techniques?

Answer: To maintain motivation, remind yourself of the benefits and track your progress. You can also create a routine by scheduling specific times for visualization, making it a regular part of your personal development practice.

Question 16: Can visualization help me overcome procrastination?

Answer: Yes, visualization can help combat procrastination by increasing your motivation and focus on your goals. By consistently visualizing your desired outcomes and the satisfaction of completing tasks, you train your mind to prioritize your objectives and take action.

Question 17: Is it possible to visualize too much or too often?

Answer: While visualization is a powerful technique, it's important to strike a balance between visualization and action. Overemphasizing visualization without taking practical steps toward your goals can hinder your progress. Use visualization to enhance your motivation and focus, but also ensure you're actively working on achieving your objectives.

Question 18: Can I use visualization to help me achieve long-term goals?

Answer: Yes, visualization is particularly effective for working toward long-term goals. By regularly visualizing your desired outcome, you maintain a clear vision of your ultimate objective, which helps you stay focused, motivated, and committed to your plan of action.

Question 19: How do I know if my visualization practice is working?

Answer: You'll likely notice a shift in your mindset, confidence, and motivation as you consistently practice visualization. You may find it easier to take action, maintain focus, and persevere through challenges, ultimately making progress toward your goals.

Question 20: Are there any resources or books you recommend for learning more about visualization techniques?

Answer: Some popular books on visualization include "Creative Visualization" by Shakti Gawain, "The Power of Your Subconscious Mind" by Joseph Murphy, and "Psycho-Cybernetics" by Maxwell Maltz. Additionally, you can find various guided visualization recordings, apps, and online resources to further explore and enhance your practice.

As an experienced life coach and motivational speaker, I understand the power of visualization in shaping a success mindset. The following techniques and tools will guide you in harnessing the power of visualization to achieve your goals and manifest the life you desire.

Guided Visualization: Find a quiet space, close your eyes, and take deep breaths. Listen to a guided visualization recording or follow a script that leads you through a detailed mental journey to your desired outcome. Engage all your senses to make the experience as vivid as possible.

Vision Board: Create a visual representation of your goals by making a collage of images, phrases, and affirmations that resonate with you. Display your vision board in a prominent place, where you can see it daily to reinforce your commitment to your goals.

Daily Visualization Practice: Dedicate 5-10 minutes each day to visualize your goals. Create a routine by setting a specific time and place for your practice. This consistency will strengthen the neural pathways in your brain, making it easier to manifest your desires.

The Mental Rehearsal Technique: Mentally rehearse a future event or task you want to excel in. Visualize yourself performing each step with ease and confidence, and feel the emotions of success as you complete the task.

Affirmations: Combine your visualization practice with positive affirmations. Speak your affirmations out loud or write them down as you visualize your desired outcomes, reinforcing your belief in your ability to achieve your goals.

Mindfulness and Meditation: Integrate mindfulness and meditation into your visualization practice. This will help you develop greater focus and mental clarity, making your visualizations more effective.

The Gratitude Technique: During your visualization sessions, take a moment to express gratitude for the achievements and positive experiences you have already manifested in your life. This will help you stay grounded and maintain a positive mindset.

Progress Tracking: Keep a journal to track your progress and document any insights or breakthroughs you experience during your visualization practice. This will help you stay motivated and committed to your goals.

Visualization Apps: Explore various visualization apps available for smartphones and tablets. These apps can provide guided visualization exercises, affirmations, and other tools to enhance your practice.

Seek Expert Guidance: Consider working with a life coach, motivational speaker, or experienced consultant to help you fine-tune your visualization techniques and provide personalized guidance on your journey to success.

Remember, the key to effective visualization lies in consistent practice and genuine belief in your ability to achieve your goals. Use these techniques and tools to unlock the power of your mind and manifest the life you've always envisioned.

Chapter 3: Nurturing Relationships for Personal Success

Enhancing Communication Skills

As an experienced life coach and motivational speaker, I know the importance of nurturing relationships for personal success. One of the most critical aspects of building and maintaining strong relationships is developing effective communication skills. In this section, we'll explore various strategies and techniques to help you enhance your communication skills and cultivate meaningful connections.

Active Listening: Active listening involves fully engaging with the speaker, giving them your undivided attention, and providing feedback to show understanding. Practice active listening by maintaining eye contact, nodding, and offering verbal affirmations.

Empathy: Put yourself in the other person's shoes and try to understand their feelings, thoughts, and perspectives. This will help you build stronger connections and foster mutual understanding in your relationships.

Nonverbal Communication: Pay attention to your body language, facial expressions, and gestures. They can convey powerful messages, even when you're not speaking. Ensure your nonverbal cues align with your verbal communication to avoid sending mixed signals.

Clear and Concise Messaging: Be clear and concise in your communication to avoid misunderstandings. Use simple language, avoid jargon, and get straight to the point.

Emotional Intelligence: Develop your emotional intelligence by recognizing and managing your emotions and those of others. This skill will help you navigate challenging conversations and foster more positive interactions.

Asking Open-Ended Questions: Encourage conversation by asking open-ended questions that require more than a simple "yes" or "no" response. This will invite deeper discussion and demonstrate your interest in the other person's thoughts and opinions.

Assertiveness: Learn to express your thoughts, feelings, and needs assertively while respecting the rights of others. This will help you maintain healthy boundaries and avoid passive or aggressive communication styles.

Conflict Resolution: Develop conflict resolution skills to address disagreements and misunderstandings in a constructive manner. Focus on finding common ground and solutions that benefit all parties involved.

Effective Feedback: Provide constructive feedback that is specific, actionable, and timely. Also, be open to receiving feedback from others and using it as an opportunity for growth.

Continuous Improvement: Embrace lifelong learning and continuously work on improving your communication skills. Attend workshops, read books, and consult with experienced professionals to stay up-to-date on best practices.

Suggested Readings:

"How to Win Friends and Influence People" by Dale Carnegie

"Crucial Conversations: Tools for Talking When Stakes Are High" by Kerry Patterson, Joseph Grenny, Ron McMillan, and Al Switzler

"The 5 Love Languages: The Secret to Love that Lasts" by Gary Chapman

Real-Life Example:

Jane, a marketing executive, struggled to communicate her ideas clearly during meetings. After attending a workshop on active listening and assertive communication, she began applying these techniques in her workplace. As a result, she noticed improved collaboration with her colleagues and increased confidence in expressing her thoughts.

In conclusion, enhancing your communication skills is essential for nurturing relationships and achieving personal success. By applying these techniques and continuously seeking improvement, you'll be on your way to building stronger connections with those around you.

Example 1:

Problem/Question: Michael, a project manager, often found himself in conflicts with his team members. He was unsure how to resolve these issues and maintain a harmonious work environment.

Answer: As an experienced life coach, I advised Michael to develop his active listening and empathy skills. By attentively listening to his team members' concerns and trying to understand their perspectives, Michael could address the root causes of conflicts and work towards constructive solutions. Over time, Michael noticed a significant improvement in team dynamics and a decrease in conflicts.

Example 2:

Problem/Question: Emily, a sales representative, had trouble connecting with her clients and closing deals. She felt that her communication style might be the issue but didn't know how to improve it.

Answer: As an experienced consultant, I suggested Emily focus on enhancing her nonverbal communication and emotional intelligence. By paying attention to her body language, facial expressions, and tone of voice, she could convey confidence and establish rapport with her clients. Additionally, by understanding her clients' emotions and needs, Emily could tailor her sales approach to address their concerns effectively. Over time, Emily saw a notable increase in her sales performance.

Example 3:

Problem/Question: Sarah and David, a married couple, were struggling to maintain open lines of communication in their relationship. They often felt unheard and misunderstood by one another.

Answer: As a motivational speaker and relationship coach, I recommended that Sarah and David practice assertiveness and ask open-ended questions. By clearly expressing their thoughts, feelings, and needs, they could avoid misunderstandings and promote honest discussions. Furthermore, asking open-ended questions would encourage deeper conversations and help them understand each other's perspectives. As they applied these techniques, Sarah and David noticed a significant improvement in their communication and overall relationship satisfaction.

By providing real-life examples of individuals facing communication challenges, readers can better understand the importance of enhancing communication skills and how to apply the strategies discussed in their lives. These stories demonstrate that with dedication and practice, anyone can improve their communication abilities and nurture relationships for personal success.

Question: How can I improve my active listening skills?

Answer: To improve your active listening skills, focus on giving your full attention to the speaker, avoid interrupting, and paraphrase what they've said to ensure understanding. Also, provide nonverbal cues such as nodding and maintaining eye contact.

Question: What are some effective ways to communicate assertively?

Answer: Communicate assertively by expressing your thoughts, feelings, and needs clearly and respectfully. Use "I" statements, maintain a confident posture, and practice assertive body language, like steady eye contact and open gestures.

Question: How can I develop my emotional intelligence?

Answer: Developing emotional intelligence involves self-awareness, self-regulation, empathy, and strong social skills. Practice reflecting on your emotions, understanding others' perspectives, and managing your reactions in various situations.

Question: How can I enhance my nonverbal communication skills?

Answer: To enhance nonverbal communication, pay attention to your body language, facial expressions, eye contact, and tone of voice. Practice maintaining open postures, conveying confidence, and mirroring others' body language to establish rapport.

Question: How can asking open-ended questions improve communication in relationships?

Answer: Open-ended questions encourage deeper conversations, as they require more than a simple "yes" or "no" response. They allow people to express themselves and provide insight into their thoughts and feelings, fostering understanding and connection.

Question: How can I address misunderstandings in communication effectively?

Answer: When misunderstandings arise, clarify your message, ask questions to understand the other person's perspective, and use active listening to ensure both parties feel heard and understood. Apologize if necessary and work towards a mutual resolution.

Question: How can I become a better communicator in my professional life?

Answer: In a professional setting, practice active listening, assertiveness, empathy, and concise communication. Tailor your communication style to your audience and be open to feedback and improvement.

Question: What is the role of feedback in effective communication?

Answer: Feedback is essential for effective communication, as it helps individuals understand how their message is received and perceived. Constructive feedback promotes growth, learning, and improvement in communication skills.

Question: How can I build rapport with someone I've just met?

Answer: To build rapport, use open body language, maintain eye contact, show genuine interest in the conversation, and find common ground. Practice active listening and ask open-ended questions to create a connection.

Question: What is the importance of empathy in communication?

Answer: Empathy allows you to understand others' perspectives and emotions, which helps foster connection, trust, and collaboration. Empathetic communication promotes healthy relationships and effective problem-solving.

Question: How can I become more adaptable in my communication style?

Answer: Observe others' communication preferences, be open to feedback, and practice adjusting your style to suit different situations, audiences, or cultural backgrounds.

Question: How can I improve my public speaking skills?

Answer: To improve public speaking skills, practice regularly, know your audience, structure your content, and focus on strong delivery, including body language, tone, and pacing. Seek feedback and embrace opportunities to speak in public.

Question: How can I overcome my fear of confrontation in communication?

Answer: Develop assertiveness, practice expressing your thoughts and feelings respectfully, and remember that confrontation can lead to positive outcomes when approached constructively.

Question: How can I communicate effectively in a virtual setting?

Answer: In a virtual setting, ensure a strong internet connection, use video when possible, maintain eye contact, speak clearly, and use visual aids to support your message. Engage your audience through questions and interactive activities.

Question: How can I handle difficult conversations with grace and confidence?

Answer: Prepare for the conversation, focus on the issue at hand, use "I" statements, listen actively, and stay calm. Be open to compromise and strive for a mutually beneficial resolution.

Question: How can I improve my written communication skills?

Answer: To improve your written communication skills, practice writing regularly, expand your vocabulary, proofread your work, and solicit feedback from others. Be clear, concise, and focused in your writing.

Question: What is the role of cultural awareness in communication?

Answer: Cultural awareness helps you understand and respect the differences in communication styles, customs, and values among various cultural backgrounds. It fosters effective communication, reduces misunderstandings, and promotes inclusiveness.

Question: How can I develop my negotiation skills?

Answer: Develop your negotiation skills by understanding the interests and needs of all parties involved, being prepared with relevant information, and practicing active listening. Work on your assertiveness, adaptability, and problem-solving skills.

Question: What is the importance of setting boundaries in communication?

Answer: Setting boundaries in communication helps protect your time, energy, and mental well-being. It fosters mutual respect, enables effective communication, and maintains healthy relationships.

Question: How can I improve my communication skills in conflict resolution?

Answer: In conflict resolution, practice active listening, empathy, assertiveness, and staying focused on the issue. Be open to finding common ground and work towards a solution that benefits all parties involved.

As a life coach, I understand the importance of communication in building and maintaining strong relationships. Here are some techniques and tools to help you enhance your communication skills:

Active Listening Exercise: Practice active listening by fully focusing on the speaker, avoiding interruptions, and providing feedback. A simple exercise is to listen to a podcast or watch a video, then summarize what you've heard without looking at any notes.

Nonverbal Communication Awareness: Observe yourself in front of a mirror or record a video while you're speaking. Pay attention to your body language, facial expressions, and gestures to identify areas for improvement.

The 3 C's Worksheet: Create a worksheet to practice the 3 C's of effective communication - Clarity, Conciseness, and Confidence. Write down a message you want to convey and work on refining it using these principles.

Emotional Intelligence Reflection: Reflect on your emotions and the emotions of others in your daily interactions. By understanding your emotional triggers and practicing empathy, you'll be better equipped to handle difficult conversations.

Assertiveness Role-Play: Role-play scenarios with a friend or family member to practice assertive communication. This will help you become more comfortable expressing your thoughts and needs while respecting the rights of others.

Improvisation Activities: Engage in improvisational exercises to develop your adaptability and quick thinking. Join a local improv group or practice with friends to improve your ability to think on your feet.

"I" Statements Technique: Learn to use "I" statements to express your feelings and thoughts without blaming or attacking others. This technique fosters open communication and reduces the chances of conflicts.

Feedback Journal: Maintain a feedback journal to document constructive criticism you receive from others. Reflect on this feedback and identify areas for improvement in your communication skills.

Reading List: Create a reading list of books, articles, and resources on communication skills. Some suggested readings include "Crucial Conversations" by Kerry Patterson and "Nonviolent Communication" by Marshall B. Rosenberg.

Guided Meditations: Practice guided meditations to increase self-awareness, manage stress, and improve your emotional intelligence. This will help you stay calm and composed during difficult conversations.

By incorporating these techniques and tools into your daily life, you'll be well on your way to enhancing your communication skills and nurturing relationships for personal success.

Strengthening Emotional Intelligence

As an experienced life coach, I know the significance of emotional intelligence in nurturing relationships and achieving personal success. Emotional intelligence involves understanding and managing your emotions and the emotions of others. It is a crucial skill that helps us navigate the complexities of our social and professional lives.

In this section, we will explore ways to strengthen your emotional intelligence, which will not only improve your relationships but also set you up for success in every aspect of your life. Let's dive into the key aspects of emotional intelligence and some practical tips on how to enhance these skills.

Self-Awareness: The foundation of emotional intelligence is self-awareness, which refers to recognizing and understanding your emotions, strengths, and weaknesses. To improve self-awareness, you can:

a. Keep a journal to record and reflect on your emotions.

b. Practice mindfulness meditation, focusing on your thoughts and feelings in the present moment.

c. Seek feedback from trusted friends, family, or colleagues to gain insights into your emotional patterns.

Self-Regulation: This involves managing your emotions and reactions, especially in challenging situations. To strengthen self-regulation, consider these strategies:

a. Identify your emotional triggers and develop coping strategies to deal with them effectively.

b. Practice relaxation techniques, such as deep breathing or progressive muscle relaxation, to calm yourself during emotional upheavals.

c. Set personal boundaries and stick to them to maintain emotional balance.

Empathy: Empathy is the ability to understand and share the feelings of others. To enhance empathy, try the following:

a. Actively listen to others, focusing on their emotions and needs.

b. Put yourself in the other person's shoes and imagine how they might feel in a given situation.

c. Develop a genuine interest in other people's lives and experiences.

Social Skills: Effective communication and relationship-building skills are essential for personal success. To improve your social skills, consider these tips:

a. Join clubs or organizations where you can interact with diverse groups of people.

b. Practice active listening, asking open-ended questions, and providing thoughtful responses.

c. Attend workshops or read books on effective communication and relationship-building.

Some suggested readings to enhance your emotional intelligence include "Emotional Intelligence" by Daniel Goleman and "The Emotional Intelligence Workbook" by Jill Dann.

By incorporating these strategies into your daily life, you'll be well on your way to strengthening your emotional intelligence and nurturing relationships for personal success. Remember that developing emotional intelligence is a lifelong process, so be patient and persistent in your efforts. The rewards will be well worth the time and energy you invest.

Story 1: Self-Awareness

Problem: Sarah often found herself in conflicts with her coworkers, without understanding why. These conflicts left her feeling emotionally drained and disconnected from her team.

Solution: After consulting with a life coach, Sarah started keeping a journal to record her emotions, thoughts, and experiences at work. By reflecting on her journal entries, she identified patterns in her behavior that contributed to the conflicts. With this newfound self-awareness, Sarah was able to address these patterns and improve her relationships at work.

Story 2: Self-Regulation

Problem: Jack's temper often got the best of him, leading to outbursts that damaged his personal and professional relationships.

Solution: Jack attended a workshop led by a motivational speaker who introduced him to relaxation techniques, such as deep breathing and progressive muscle relaxation. By practicing these techniques regularly, Jack learned to manage his emotions more effectively, resulting in better relationships and fewer conflicts.

Story 3: Empathy

Problem: Amanda's lack of empathy made it difficult for her to connect with others and build meaningful relationships.

Solution: To improve her empathy, Amanda began volunteering at a local shelter, where she interacted with people from diverse backgrounds and experiences. This exposure helped her develop a genuine interest in others' lives and feelings, leading to stronger connections and more fulfilling relationships.

Story 4: Social Skills

Problem: Tom struggled with social anxiety and found it challenging to communicate effectively with others.

Solution: To overcome his social anxiety, Tom joined a local Toastmasters club, where he practiced public speaking and engaged in group discussions with diverse individuals. Over time, Tom improved his communication skills and gained the confidence to build strong relationships in his personal and professional life.

By learning from these real-life examples, you can see the transformative power of strengthening emotional intelligence. Embrace the strategies outlined in this chapter to enhance your self-awareness, self-regulation, empathy, and social skills, and watch your relationships and personal success flourish. Remember, the journey to emotional intelligence is a continuous one - stay committed to your growth and be open to learning from the experiences of others.

Q: What are the four main components of emotional intelligence?

A: The four main components are self-awareness, self-regulation, empathy, and social skills.

Q: How can journaling help improve self-awareness?

A: Journaling encourages reflection on emotions, thoughts, and experiences, allowing individuals to identify patterns in their behavior and better understand themselves.

Q: Can you suggest a relaxation technique for improving self-regulation?

A: Deep breathing exercises or progressive muscle relaxation are effective techniques for managing emotions and improving self-regulation.

Q: What is a practical way to develop empathy?

A: Volunteering or engaging with people from diverse backgrounds can help cultivate empathy and understanding.

Q: How can joining a public speaking club help improve social skills?

A: Public speaking clubs, such as Toastmasters, provide opportunities for individuals to practice communication and engage in group discussions, which can help overcome social anxiety and develop confidence.

Q: How does emotional intelligence impact personal relationships?

A: Emotional intelligence allows individuals to better understand and manage their emotions, empathize with others, and communicate effectively, leading to stronger and more fulfilling relationships.

Q: Can emotional intelligence be learned or improved?

A: Yes, emotional intelligence can be developed and enhanced through practice, self-reflection, and learning from experiences.

Q: Why is self-awareness essential for personal success?

A: Self-awareness helps individuals recognize their strengths, weaknesses, and behavioral patterns, enabling them to make informed decisions and adapt to various situations effectively.

Q: How can one practice empathy in daily life?

A: Actively listening to others, asking open-ended questions, and being genuinely interested in their feelings and experiences can help practice empathy in everyday interactions.

Q: What role does emotional intelligence play in conflict resolution?

A: Emotional intelligence enables individuals to manage their emotions, empathize with others, and communicate effectively, making it easier to resolve conflicts and find solutions that satisfy all parties involved.

Q: How can someone with social anxiety work on improving their emotional intelligence?

A: Gradual exposure to social situations, practicing relaxation techniques, and engaging in activities that promote communication and interaction can help individuals with social anxiety develop emotional intelligence.

Q: Can mindfulness meditation help improve emotional intelligence?

A: Yes, mindfulness meditation promotes self-awareness, emotional regulation, and empathy, contributing to the development of emotional intelligence.

Q: How does emotional intelligence impact professional success?

A: Emotional intelligence helps individuals navigate interpersonal relationships, manage stress, and adapt to change, leading to increased productivity, better decision-making, and overall professional success.

Q: What is the connection between emotional intelligence and mental health?

A: Strong emotional intelligence can help individuals manage their emotions effectively, cope with stress, and build resilience, which are essential for maintaining good mental health.

Q: How can someone identify areas for improvement in their emotional intelligence?

A: Self-reflection, seeking feedback from others, and assessing one's reactions to various situations can help individuals identify areas for improvement in their emotional intelligence.

Q: Can role-playing exercises help develop emotional intelligence?

A: Yes, role-playing exercises allow individuals to practice empathy, communication, and emotional regulation in a controlled environment, helping to enhance emotional intelligence.

Q: How can a life coach or consultant help improve emotional intelligence?

A: A life coach or consultant can provide guidance, strategies, and tools to help individuals develop self-awareness, self-regulation, empathy, and social skills, ultimately improving emotional intelligence.

Q: What is the importance of setting boundaries in personal relationships?

A: Setting boundaries helps individuals maintain

Technique 1: Emotional Journaling

To strengthen your emotional intelligence, start by journaling your emotions daily. Write down your feelings, thoughts, and reactions to events, people, and situations. This practice encourages self-awareness and helps you identify patterns in your emotional responses.

Technique 2: Active Listening

Develop the habit of active listening by fully focusing on the person speaking, without interrupting or forming judgments. This will improve your empathy and understanding of others' emotions.

Technique 3: Emotional Vocabulary Expansion

Expand your emotional vocabulary by learning new words and phrases to describe feelings accurately. This will allow you to articulate and understand your emotions better, contributing to improved emotional intelligence.

Technique 4: Mindfulness Meditation

Practice mindfulness meditation to heighten your self-awareness and emotional regulation. Through mindfulness, you can learn to recognize your emotions and respond to them in a balanced manner.

Technique 5: Seeking Feedback

Seek constructive feedback from trusted individuals to identify areas of improvement in your emotional understanding and interpersonal skills. Use this feedback to develop a plan for growth.

Technique 6: Empathy Exercises

Participate in empathy-building exercises, such as role-playing or reading fiction, to understand diverse perspectives and emotions better. These activities help develop empathy, a crucial component of emotional intelligence.

Technique 7: Conflict Resolution Skills

Learn and practice conflict resolution skills to manage disagreements effectively and foster positive relationships. This will improve your emotional regulation and interpersonal skills.

Technique 8: Self-Compassion Practice

Cultivate self-compassion by treating yourself with kindness and understanding during difficult times. This practice enhances emotional regulation and resilience.

Technique 9: Emotional Regulation Techniques

Develop emotional regulation techniques, such as deep breathing, progressive muscle relaxation, or visualization, to manage emotional responses effectively.

Technique 10: Continuous Learning

Commit to continuous learning and growth in emotional intelligence. Attend workshops, read books, or work with a life coach to further develop your understanding of emotions and enhance your interpersonal skills.

Building and Maintaining Supportive Networks

Building and maintaining a supportive network is crucial for personal success. Your network should include people who can offer guidance, encouragement, and resources to help you achieve your goals. In this section, we will discuss the steps to build and maintain a strong network, along with real-life examples and questions to guide your journey.

Step 1: Identify Your Goals and Values

Before building your network, it's essential to know your goals and values. This will allow you to seek connections with people who share your objectives and principles, fostering a more productive and supportive environment.

Step 2: Attend Networking Events

Join networking events, conferences, and workshops relevant to your interests and goals. These gatherings provide opportunities to meet like-minded individuals and expand your network.

Step 3: Leverage Social Media

Utilize social media platforms such as LinkedIn, Twitter, and Facebook to connect with professionals and influencers in your field. Share valuable content and engage in conversations to establish rapport and credibility.

Step 4: Nurture Existing Relationships

Maintain connections with friends, family, and colleagues who have supported you throughout your journey. Express gratitude for their assistance and invest time in these relationships to strengthen your support system.

Step 5: Offer Help and Support to Others

Be proactive in providing help and support to those in your network. By doing so, you demonstrate your value and commitment to fostering mutually beneficial relationships.

Step 6: Follow-up and Stay Connected

Regularly follow up with your connections to stay updated on their progress and achievements. This consistent communication builds trust and strengthens the bond between you and your network.

Real-life example:

Sarah, an aspiring entrepreneur, wanted to expand her network to gain insights and resources for her business. She attended local networking events and joined online forums relevant to her field. By engaging with others and offering support, Sarah was able to establish a strong network of professionals who provided valuable advice and resources for her business.

Question:

How can you leverage social media to expand your professional network?

Answer:

You can leverage social media by connecting with professionals and influencers in your field, engaging in relevant conversations, sharing valuable content, and participating in online networking events.

Suggested Readings:

"Never Eat Alone" by Keith Ferrazzi

"The Tipping Point" by Malcolm Gladwell

"How to Win Friends and Influence People" by Dale Carnegie

Remember, building and maintaining a supportive network takes time and effort. By following these steps and adopting a proactive, genuine approach to networking, you can establish connections that will contribute to your personal success.

Real-life Story 1:

Mike, a talented software developer, found himself struggling to find new opportunities in his field. He knew he had the skills but lacked connections in the industry.

Problem/Question: How could Mike expand his network and increase his chances of finding new opportunities?

Answer: Mike decided to attend local tech meetups, conferences, and workshops related to his field. He also joined online forums and engaged with others on social media platforms like LinkedIn and Twitter. By actively participating in discussions, sharing his expertise, and offering help to others, Mike was able to build a strong network of professionals who eventually led him to new career opportunities.

Real-life Story 2:

Emma, a young professional, was looking for a mentor to guide her through her career development. Despite having numerous connections on LinkedIn, she couldn't find someone willing to mentor her.

Problem/Question: How could Emma identify a suitable mentor within her network and establish a mentoring relationship?

Answer: Emma decided to analyze her LinkedIn connections and identify professionals who had experience in her field and shared her values. She reached out to them with personalized messages, expressing her admiration for their work and her desire to learn from their expertise. After several attempts, she found a mentor who appreciated her initiative and was willing to provide guidance and support.

Real-life Story 3:

Carlos, a small business owner, was struggling to keep his company afloat. He needed support from a network of entrepreneurs who could provide guidance and resources.

Problem/Question: How could Carlos build a supportive network to help his business thrive?

Answer: Carlos started attending local networking events and joined entrepreneur-focused online communities. He made an effort to engage with other business owners, offering his own experience and support. By actively participating and providing value to others, Carlos built a network of entrepreneurs who offered him valuable advice, resources, and connections to help his business succeed.

Question: How do I begin building a supportive network?

Answer: Start by attending networking events, joining online communities, and connecting with people in your field. Be genuine, open-minded, and willing to engage in meaningful conversations.

Question: How can I maintain the connections I've made?

Answer: Consistently communicate with your connections, share valuable information, and offer help when possible. This will foster strong relationships and make your network more supportive.

Question: How can I identify potential mentors within my network?

Answer: Look for individuals who have experience in your field, share your values, and have a willingness to share their knowledge.

Question: How do I know if a networking event is worth attending?

Answer: Research the event, consider the attendees, and evaluate whether the event aligns with your goals and interests.

Question: How can I make a positive first impression when networking?

Answer: Be genuine, approachable, and actively listen to others. Show genuine interest in their experiences and ideas.

Question: What is the best way to ask for help from my network?

Answer: Be specific about your needs, explain the situation, and express your gratitude for any help offered.

Question: How can I offer value to my network?

Answer: Share your expertise, provide support, and connect people within your network who may benefit from each other's resources.

Question: What is the best way to follow up with someone I've met at a networking event?

Answer: Send a personalized message expressing your appreciation for their time and highlighting any common interests or discussion points.

Question: How do I maintain my network while balancing my professional and personal life?

Answer: Prioritize your connections, schedule regular check-ins, and utilize digital tools to stay connected.

Question: Can introverts be successful at networking?

Answer: Absolutely. Introverts can excel at networking by focusing on meaningful, one-on-one conversations and building deep connections.

Question: How do I approach someone at a networking event if I feel nervous?

Answer: Take a deep breath, remind yourself of your goals, and initiate a conversation with a friendly smile and open-ended questions.

Question: How can I diversify my network?

Answer: Attend events in different industries, join various online communities, and connect with people from diverse backgrounds.

Question: How do I know if my network is supportive?

Answer: A supportive network is one where members actively engage, share resources, offer guidance, and provide encouragement.

Question: How can I improve my networking skills?

Answer: Practice active listening, enhance your communication skills, and develop a genuine interest in others.

Question: What if I don't have a large network?

Answer: Focus on the quality of your connections rather than the quantity. Even a small network can provide valuable support.

Question: How do I handle rejection in networking?

Answer: Stay positive, learn from the experience, and continue building connections.

Question: How often should I attend networking events?

Answer: Determine a schedule that works for you, whether that's attending events monthly, quarterly, or yearly.

Question: How can I leverage my network to advance my career?

Answer: Seek guidance from mentors, ask for introductions to potential employers, and stay updated on industry trends.

Question: Can I network with people outside of my industry?

Answer: Yes, networking with people from different industries can provide fresh perspectives and potential collaboration opportunities.

Question: What are some common networking mistakes to avoid?

Answer: Avoid dominating conversations, focusing solely on your own interests, and neglecting to follow up with connections.

As an experienced life coach and motivational speaker, I understand the power of building and maintaining supportive networks. In this section, we will discuss practical techniques and tools to help you create strong connections that can propel you towards personal and professional success. By following these steps, you will be better equipped to navigate the complex world of networking.

The Mindful Networker's Journal: Start by keeping a journal to track your networking efforts. Reflect on your experiences, the people you've met, and the lessons you've learned. This mindfulness practice will help you identify patterns and areas for improvement, making you a more effective networker.

The Relationship Mapping Technique: Create a visual representation of your network by listing the names of your connections and drawing lines to indicate the relationships between them. This will help you identify gaps in your network and opportunities for introductions between connections.

The Elevator Pitch: Develop a concise and compelling description of who you are and what you do. Practice delivering your elevator pitch with confidence, ensuring you leave a lasting impression on those you meet.

Active Listening Exercises: To strengthen your communication skills, practice active listening exercises with a friend or family member. Focus on truly understanding their perspective and asking thoughtful questions to demonstrate genuine interest in their experiences.

The Networking Action Plan: Create a plan outlining your networking goals, including the types of connections you want to make, events to attend, and strategies for following up with new contacts. Revisit and update your plan regularly to stay focused and motivated.

The 5-3-1 Method: After each networking event, identify five new connections you made, three valuable insights you gained, and one action step you will take to follow up with your new contacts.

The Gratitude Practice: Express gratitude to your network by sending thoughtful messages, offering assistance, or sharing valuable resources. Cultivating a spirit of gratitude not only strengthens your relationships but also reinforces a positive mindset.

Digital Networking Mastery: Leverage social media platforms like LinkedIn to expand your network and maintain connections. Keep your profile up-to-date and actively engage with others through comments, messages, and sharing content.

The Networking Book Club: Start a book club focused on personal and professional development with members of your network. This will facilitate deeper connections, provide opportunities for shared learning, and demonstrate your commitment to growth.

The Networking Accountability Partner: Find someone within your network who shares your networking goals and hold each other accountable for taking action. Regular check-ins and encouragement will help you both stay on track and motivated.

By incorporating these techniques and tools into your networking efforts, you will foster meaningful connections and create a supportive network that contributes to your personal success. Remember, as an experienced life coach and motivational speaker, my goal is to empower you to take charge of your networking journey and forge the relationships that will propel you towards your dreams.

Conflict Resolution and Negotiation Skills

As an experienced life coach and motivational speaker, I know that conflicts are a natural part of human relationships. Learning to navigate disagreements and negotiate effectively is essential for personal and professional success. In this section, I will provide practical advice on conflict resolution and negotiation skills, supported by real-life examples and a wealth of resources to help you become a more adept communicator.

Understanding Conflict: Recognize that conflict is a natural part of human interaction, and it can lead to growth, innovation, and stronger relationships when managed effectively. Start by educating yourself on the sources of conflict, such as differing values, expectations, or communication styles.

Embrace Empathy: Practice empathy by putting yourself in the other person's shoes and genuinely trying to understand their perspective. This approach can help defuse tension and pave the way for constructive dialogue.

Active Listening: Focus on listening attentively to the other party, without interrupting or formulating your response while they speak. Active listening demonstrates respect and can lead to a deeper understanding of the issue at hand.

Use "I" Statements: Express your thoughts and feelings using "I" statements rather than accusatory language. This can help prevent the other party from becoming defensive and promote a more open, honest discussion.

Seek Win-Win Solutions: When negotiating, aim for mutually beneficial outcomes that address the needs and concerns of both parties. Collaboration, rather than competition, can lead to more satisfying and sustainable resolutions.

Practice Assertiveness: Stand up for your needs and wants while maintaining respect for the other person's perspective. Assertiveness involves clear, confident communication and a willingness to compromise when necessary.

Manage Emotions: Learn to recognize and manage your emotions during conflict, avoiding impulsive reactions that can escalate tension. Techniques such as deep breathing, self-awareness, and emotional regulation can help you stay calm and focused.

Be Open to Feedback: Embrace constructive criticism and feedback as an opportunity for personal growth. Reflect on the lessons learned and commit to making improvements in the future.

Reflect and Learn: After resolving a conflict or negotiation, take time to reflect on the process, the outcome, and what you learned. Consider how you can apply these insights to future conflicts and negotiations.

Develop Your Skills: Continually invest in your conflict resolution and negotiation skills through workshops, books, and online courses. Some recommended resources include "Getting to Yes" by Roger Fisher and William Ury, and "Crucial Conversations" by Kerry Patterson and Joseph Grenny.

By incorporating these strategies into your communication toolkit, you will be better equipped to navigate conflicts and negotiations with confidence and skill. As an experienced life coach and motivational speaker, my aim is to empower you to foster healthy, productive relationships that contribute to your personal success. Embrace these lessons, and remember that every conflict presents an opportunity for growth and learning.

Example 1: Negotiating a Raise at Work

Problem: Jane, a hardworking marketing specialist, felt underpaid compared to her colleagues. She decided to request a raise from her boss but was unsure how to approach the situation.

Solution: Jane prepared for the negotiation by researching industry salary standards and collecting evidence of her accomplishments. During the meeting, she used active listening and empathy to understand her boss's perspective. She communicated her needs assertively and presented a win-win solution that acknowledged both her contributions and the company's budget constraints. Ultimately, her boss agreed to a raise and a performance-based bonus structure.

Example 2: Resolving a Dispute Between Roommates

Problem: Tom and Jerry, college roommates, constantly argued about chores and noise levels. Tension escalated, and their friendship began to suffer.

Solution: They agreed to sit down and discuss their issues openly. Tom practiced active listening, allowing Jerry to express his concerns without interruption. They both used "I" statements to avoid accusations and focused on finding a mutually agreeable solution. In the end, they created a chore schedule and established quiet hours, which resolved their conflicts and strengthened their friendship.

Example 3: Dealing with a Difficult Co-worker

Problem: Sarah was frustrated with her co-worker, Kevin, who often took credit for her ideas and undermined her authority in front of their team.

Solution: Sarah sought advice from an experienced consultant who encouraged her to address the issue directly with Kevin. During a private conversation, she managed her emotions and used assertive communication to express her concerns. Kevin was initially defensive, but Sarah's empathetic approach allowed them to find common ground. They agreed to be more collaborative and respectful moving forward.

Example 4: Settling a Family Dispute

Problem: Siblings Mia and Sam were at odds over the care of their elderly mother, with each believing the other wasn't doing enough to help.

Solution: They enlisted the help of a life coach skilled in conflict resolution to facilitate a family meeting. The coach encouraged them to embrace empathy, practice active listening, and share their feelings using "I" statements. The siblings realized they both wanted the best for their mother and agreed to a care plan that addressed their concerns and balanced their responsibilities.

By implementing the conflict resolution and negotiation skills discussed in this section, individuals like Jane, Tom, Jerry, Sarah, Mia, and Sam were able to overcome their challenges and nurture healthier, more productive relationships.

Q: How can I approach a conflict without making it worse?

A: Use active listening, empathize with the other person's perspective, and avoid accusatory language. Focus on finding a mutually agreeable solution.

Q: What is the role of empathy in conflict resolution?

A: Empathy allows you to understand the other person's feelings and needs, fostering a more open and collaborative dialogue.

Q: How can I stay calm during a heated conflict?

A: Practice deep breathing, maintain a neutral tone, and remind yourself to focus on finding a solution instead of winning the argument.

Q: What are some effective negotiation strategies?

A: Prepare beforehand, establish rapport, listen actively, and present a win-win solution.

Q: How do I avoid being too passive or aggressive in a negotiation?

A: Practice assertive communication, expressing your needs and opinions respectfully while considering the other party's perspective.

Q: How can I become a better active listener?

A: Focus on the speaker, avoid interrupting, and provide feedback to show understanding.

Q: What are "I" statements, and how can they help in conflict resolution?

A: "I" statements express feelings and concerns without blaming the other person, making communication more effective and less confrontational.

Q: How can I create a win-win solution in a negotiation?

A: Identify common goals, explore alternative solutions, and collaborate to find a mutually beneficial outcome.

Q: How can I handle a conflict with a co-worker who constantly undermines me?

A: Address the issue privately, manage your emotions, and use assertive communication to express your concerns and seek resolution.

Q: What is the importance of establishing rapport during a negotiation?

A: Rapport builds trust and facilitates open communication, increasing the likelihood of a successful outcome.

Q: How can I prevent conflicts from escalating?

A: Address issues early, maintain open communication, and practice empathy to resolve disagreements before they become unmanageable.

Q: How do I negotiate a raise at work?

A: Research industry standards, prepare evidence of your accomplishments, and present a win-win solution that acknowledges your contributions and the company's needs.

Q: How can I improve my assertive communication skills?

A: Practice expressing your thoughts and needs respectfully, maintain eye contact, and use a calm and confident tone of voice.

Q: What role does emotional intelligence play in conflict resolution?

A: Emotional intelligence helps you manage your emotions, empathize with others, and communicate effectively, all of which are essential in resolving conflicts.

Q: How can I resolve disputes within my family?

A: Establish open communication, practice active listening, and collaborate to find a solution that addresses everyone's concerns.

Q: How can I prepare for a negotiation?

A: Research the topic, gather relevant information, identify your goals, and develop alternative solutions.

Q: What are some common barriers to effective communication in conflicts?

A: Interrupting, defensive behavior, negative body language, and accusatory language can hinder productive dialogue.

Q: How can I resolve conflicts with roommates or friends?

A: Establish open communication, use active listening, and collaborate to find mutually agreeable solutions.

Q: How can I apply conflict resolution and negotiation skills to my personal relationships?

A: Practice empathy, assertive communication, and active listening to address and resolve disagreements with loved ones.

Q: How can I ensure that a conflict is truly resolved?

A: Verify that both parties feel heard and understood, establish a plan for addressing the issue, and maintain open communication

As a life coach and motivational speaker, I have found that conflict resolution and negotiation skills are essential for personal and professional success. In this section, I will guide you through practical techniques and exercises to help you develop and refine these skills.

Active Listening Exercise:

Find a partner, and take turns sharing a personal story or issue for two minutes each. While one person speaks, the other should practice active listening by maintaining eye contact, nodding, and occasionally paraphrasing the speaker's words. Afterward, discuss how it felt to be both the speaker and the listener.

The Assertive Communication Technique:

Practice expressing your thoughts and feelings using "I" statements. For example, instead of saying, "You always interrupt me," try, "I feel frustrated when I'm interrupted during conversations." This approach reduces blame and helps create a more open dialogue.

Win-Win Solution Worksheet:

Create a worksheet with two columns labeled "My Interests" and "Their Interests." List your priorities and the other party's priorities in the respective columns. Identify common ground, brainstorm solutions, and evaluate each idea for mutual benefit.

The Empathy Map:

Draw a simple face on a piece of paper. Around the face, write down what the other person might be thinking, feeling, and needing during a conflict. This exercise helps to cultivate empathy and understanding.

Role-Playing Scenarios:

With a partner, role-play various conflict or negotiation scenarios. This practice will help you develop flexibility and adaptability in real-life situations.

Mindfulness Meditation for Conflict Resolution:

Spend 10 minutes a day practicing mindfulness meditation. Focus on your breath and observe your thoughts and emotions without judgment. This practice can help you remain calm and centered during conflicts.

The Feedback Loop:

After resolving a conflict or completing a negotiation, reflect on the experience. Identify what went well, what could have been better, and what you can do differently next time. This self-assessment will help you grow and improve your skills.

The Appreciation Journal:

Keep a journal where you record instances where you successfully resolved conflicts or negotiated effectively. Review your journal regularly to reinforce your progress and boost your confidence.

These tools and exercises, when practiced consistently, will help you develop your conflict resolution and negotiation skills. Remember, as an experienced consultant and motivational speaker, I encourage you to commit to the process and embrace growth. Your personal and professional success depends on your ability to navigate relationships and resolve conflicts effectively.

Chapter 4: Achieving Balance in Life

Managing Stress and Overcoming Burnout

As an experienced life coach and motivational speaker, I understand the importance of achieving balance in life. In today's fast-paced world, stress and burnout are common challenges that can hinder our ability to live fulfilling lives. In this section, I will provide you with practical advice and strategies to manage stress and overcome burnout.

Stress is a natural response to the challenges we face in our daily lives. However, prolonged stress can lead to burnout, a state of emotional, physical, and mental exhaustion. To prevent burnout and maintain balance, it is crucial to develop effective stress management techniques and self-care habits.

Identify stressors: Start by recognizing the factors that contribute to your stress. Keep a stress journal to track your daily experiences, noting the situations, people, or tasks that cause stress. This information will help you develop targeted strategies to manage and reduce stress.

Prioritize self-care: Make time for activities that rejuvenate and relax you. This may include exercise, meditation, spending time with loved ones, or pursuing hobbies. Regular self-care helps maintain emotional and mental well-being, preventing burnout.

Set boundaries: Learn to say "no" when necessary and establish clear boundaries between your personal and professional life. Avoid overcommitting and prioritize tasks based on their importance.

Develop time management skills: Effective time management helps reduce stress by ensuring you allocate sufficient time to complete tasks. Create a daily schedule, set realistic goals, and break large tasks into smaller, manageable steps.

Seek support: Reach out to friends, family, or professional counselors for guidance and support. Sharing your experiences and seeking advice can help alleviate stress and provide new perspectives on challenging situations.

Practice mindfulness: Mindfulness techniques, such as meditation and deep breathing exercises, help reduce stress by promoting relaxation and mental clarity. Set aside a few minutes each day to practice mindfulness and develop greater self-awareness.

Suggested Readings:

"The Art of Stress-Free Living" by Dr. Brian Luke Seaward

"Burnout: The Secret to Unlocking the Stress Cycle" by Emily Nagoski and Amelia Nagoski

Real-life example:

Sarah, a marketing manager, found herself constantly overwhelmed by work demands, leaving her with little time for personal pursuits. As a result, she began to experience burnout symptoms, such as fatigue, irritability, and difficulty concentrating.

To manage her stress and overcome burnout, Sarah began implementing the following strategies:

Identifying stressors: Sarah pinpointed the factors contributing to her stress, such as unrealistic deadlines and excessive workload.

Prioritizing self-care: She committed to regular exercise and made time for socializing with friends.

Setting boundaries: Sarah learned to delegate tasks and communicate her limits to colleagues and supervisors.

Developing time management skills: She started using a daily planner and prioritized tasks based on importance.

Seeking support: Sarah joined a support group for professionals experiencing burnout and sought guidance from a life coach.

Practicing mindfulness: She incorporated meditation and deep breathing exercises into her daily routine.

By implementing these strategies, Sarah successfully managed her stress and overcame burnout, allowing her to achieve a more balanced and fulfilling life.

Example 1:

Problem: Tom, a dedicated teacher, found himself constantly juggling lesson planning, grading, and extracurricular activities. He felt overwhelmed, leading to chronic stress and, ultimately, burnout.

Solution: Tom sought the help of a life coach, who suggested he focus on self-care and time management. By setting aside time for himself, delegating tasks, and adopting better time management techniques, Tom regained his passion for teaching and achieved a healthier work-life balance.

Example 2:

Problem: Maria, a single mother, struggled to balance her demanding job with raising her two children. She felt drained and stressed, which negatively impacted her relationships and overall well-being.

Solution: Maria reached out to a support group for single parents, where she learned new coping strategies and received emotional support from others in similar situations. She also began practicing mindfulness and setting boundaries, which helped her reduce stress and find balance in her life.

Example 3:

Problem: Raj, a successful entrepreneur, experienced high levels of stress as his business grew. His work consumed all his time, causing him to neglect his personal life and health.

Solution: Raj consulted an experienced consultant who advised him to delegate tasks, prioritize self-care, and set boundaries between work and personal life. By implementing these strategies, Raj managed to reduce stress, avoid burnout, and achieve a more balanced lifestyle.

Example 4:

Problem: Lisa, a healthcare professional, found herself facing compassion fatigue as she cared for her patients during a global health crisis. Her high levels of stress led her to experience symptoms of burnout.

Solution: Lisa turned to an online course on stress management and burnout prevention. She learned to identify her stressors, practice self-care, and develop better time management skills. As a result, she overcame burnout and rediscovered her passion for helping others.

Example 5:

Problem: Michael, an aspiring writer, was struggling to balance his day job, writing projects, and family life. His stress levels were high, and he began to feel burned out.

Solution: Michael attended a motivational speaker's seminar on achieving balance in life. Inspired by the speaker's advice, he committed to setting clear boundaries, prioritizing self-care, and seeking support from friends and family. By implementing these changes, Michael found the balance he needed to thrive in all aspects of his life.

Q: What are some common signs of burnout?

A: Signs of burnout may include chronic fatigue, irritability, loss of motivation, reduced productivity, feelings of detachment, and physical symptoms like headaches or insomnia.

Q: How can I identify my stress triggers?

A: To identify stress triggers, maintain a journal to record situations or events that cause stress, noting how you react to them. This practice can help reveal patterns and highlight specific stressors.

Q: What are some effective strategies for managing stress?

A: Strategies for managing stress include regular exercise, meditation, deep breathing exercises, setting boundaries, seeking social support, prioritizing self-care, and practicing good sleep hygiene.

Q: How can I create a healthier work-life balance?

A: To create a healthier work-life balance, set clear boundaries between work and personal life, delegate tasks when possible, prioritize self-care, and make time for hobbies, friends, and family.

Q: What role does self-care play in preventing burnout?

A: Self-care is essential for maintaining mental, emotional, and physical well-being. By prioritizing self-care, you can replenish your energy, reduce stress, and improve resilience, helping to prevent burnout.

Q: How can I implement mindfulness techniques to manage stress?

A: Practice mindfulness by focusing on the present moment, without judgment. You can incorporate mindfulness into your daily life through meditation, deep breathing exercises, or simply being fully present during everyday activities.

Q: Can delegating tasks help prevent burnout?

A: Yes, delegating tasks can help prevent burnout by reducing your workload, allowing you to focus on essential tasks and avoid becoming overwhelmed.

Q: How important is setting boundaries for stress management?

A: Setting boundaries is crucial for stress management, as it helps you protect your time and energy, allowing you to prioritize self-care and maintain a healthier work-life balance.

Q: What role does time management play in preventing burnout?

A: Effective time management helps you stay organized, prioritize tasks, and allocate time for self-care, reducing stress and minimizing the risk of burnout.

Q: How can seeking support from others help manage stress and prevent burnout?

A: Seeking support from friends, family, or professional networks can provide emotional support, practical advice, and encouragement, which can help reduce stress and prevent burnout.

Q: How can I recognize when I need a break to prevent burnout?

A: Listen to your body and mind for signs of fatigue, irritability, or reduced productivity. When you notice these symptoms, take a break to rest, recharge, and reevaluate your priorities.

Q: How can hobbies and interests contribute to stress management?

A: Engaging in hobbies and interests can provide a mental break, help you relax, and promote overall well-being, which contributes to stress management.

Q: Are regular breaks important for managing stress?

A: Yes, regular breaks are essential for maintaining focus, productivity, and mental well-being, helping to manage stress and prevent burnout.

Q: How can I create a self-care routine to manage stress?

A: Create a self-care routine by scheduling time for activities that nourish your mind, body, and spirit, such as exercise, relaxation techniques, hobbies, and socializing with loved ones.

Q: How can practicing gratitude help in managing stress?

A: Practicing gratitude can help shift your focus from what is lacking to what is going well, promoting a positive mindset and reducing stress.

Q: What are some healthy ways to cope with stress

A: Healthy ways to cope with stress include regular exercise, deep breathing exercises, meditation, spending time in nature, engaging in hobbies, seeking social support, and maintaining a balanced diet.

Q: How can I set realistic expectations for myself to avoid burnout?

A: To set realistic expectations, break down goals into smaller, achievable tasks, prioritize what's most important, and be honest with yourself about your limitations. Remember that it's okay to ask for help when needed.

Q: Can making lifestyle changes, such as diet and exercise, help manage stress?

A: Yes, making lifestyle changes like maintaining a balanced diet and engaging in regular physical activity can improve overall well-being, increase resilience, and help manage stress.

Q: How can I effectively communicate my needs and boundaries to others to prevent burnout?

A: To effectively communicate your needs and boundaries, practice assertive communication, clearly express your feelings and expectations, and be open to feedback from others. Don't be afraid to say no when necessary.

Q: How can I maintain a positive mindset in high-stress situations?

A: To maintain a positive mindset, practice gratitude, focus on solutions instead of problems, surround yourself with positive influences, and utilize stress-reducing techniques such as deep breathing or visualization.

Technique 1: The Stress Inventory Worksheet

As a Life Coach, I recommend creating a Stress Inventory Worksheet to help you identify stressors in your life. List all the factors causing stress, rate their severity, and brainstorm possible solutions to each. This exercise helps you gain clarity and develop an action plan to reduce stress.

Technique 2: The 4-7-8 Breathing Exercise

Practice the 4-7-8 Breathing Exercise to calm your mind and body. Inhale for 4 seconds, hold the breath for 7 seconds, and exhale for 8 seconds. Repeat this cycle 4 times. This breathing technique can reduce anxiety and promote relaxation.

Technique 3: Time Blocking

As an Experienced Consultant, I suggest using the Time Blocking method to manage your daily tasks and reduce burnout. Divide your day into blocks dedicated to specific tasks or categories, such as work, family, exercise, or self-care. This approach ensures a balanced schedule and prevents you from becoming overwhelmed.

Technique 4: Gratitude Journal

Maintain a Gratitude Journal to remind yourself of the positive aspects of your life. Each day, write down three things you're grateful for. This practice can shift your focus from stress to appreciation, improving your overall well-being.

Technique 5: Mindfulness Meditation

Incorporate mindfulness meditation into your daily routine to reduce stress and achieve balance. Set aside 10-20 minutes to sit quietly, focusing on your breath and observing your thoughts without judgment. This practice helps you stay present and manage stress more effectively.

Technique 6: The "NO" Matrix

As a Motivational Speaker, I encourage you to create a "NO" Matrix to set boundaries and protect your time and energy. The matrix is a simple table with four quadrants: "Say Yes," "Negotiate," "Delegate," and "Say No." Categorize requests or tasks into these quadrants based on your priorities, values, and commitments. This tool can help you make informed decisions and prevent burnout.

Technique 7: Progressive Muscle Relaxation

Practice Progressive Muscle Relaxation to release physical tension and manage stress. Starting with your feet and working your way up, tense each muscle group for 5 seconds and then release for 15 seconds. This technique can help you become more aware of your body's stress signals and promote relaxation.

Incorporating Mindfulness and Meditation

Mindfulness and meditation are powerful tools that can transform your life and help you achieve balance. As an experienced Life Coach, I've seen countless clients benefit from these practices, leading to reduced stress, improved mental clarity, and enhanced emotional resilience.

Mindfulness is the practice of being present and fully engaged in the moment, without judgment or distraction. This state of awareness can be cultivated through various techniques such as mindful breathing, body scans, and daily mindfulness exercises. Meditation, on the other hand, is a focused practice that trains the mind to achieve greater concentration, relaxation, and self-awareness.

To begin incorporating mindfulness and meditation into your life, follow these steps:

Choose a practice that resonates with you: There are many forms of meditation and mindfulness practices available. Experiment with different techniques to find one that suits your needs and preferences. Some popular options include mindfulness-based stress reduction (MBSR), loving-kindness meditation, and transcendental meditation.

Create a dedicated space: Set aside a quiet, comfortable space in your home where you can practice mindfulness and meditation without distractions. This will help you establish a routine and signal to your mind that it's time to relax and focus.

Set aside time: Consistency is key when developing a mindfulness or meditation practice. Choose a specific time each day to practice, whether it's in the morning, during lunch, or before bedtime. Aim for at least 10-20 minutes daily to experience the full benefits.

Be patient and persistent: Mindfulness and meditation are skills that require practice and patience. You may experience challenges or setbacks, but remember that it's normal. Stay committed, and you'll see progress over time.

Real-life Example:

Sarah, a busy executive, struggled with chronic stress and anxiety. As an experienced Consultant, I suggested she incorporate mindfulness and meditation into her daily routine. Sarah began practicing MBSR for 20 minutes each morning and found that her stress levels decreased significantly, her focus improved, and she became more resilient in the face of challenges.

Suggested Readings:

"The Miracle of Mindfulness" by Thich Nhat Hanh

"Wherever You Go, There You Are" by Jon Kabat-Zinn

"10% Happier" by Dan Harris

Questions and Answers:

Q: How long does it take to see the benefits of mindfulness and meditation?

A: While individual experiences may vary, many people notice positive changes in their well-being after just a few weeks of consistent practice.

Q: Can I practice mindfulness and meditation if I have never done it before?

A: Absolutely! Mindfulness and meditation are suitable for people of all experience levels. Start with simple techniques and gradually build your practice over time.

Q: How can I maintain my mindfulness and meditation practice when life gets busy?

A: Prioritize your practice by scheduling it into your daily routine. Remember that even short sessions can be beneficial, so don't hesitate to practice for just a few minutes when time is limited.

Remember, as a motivational speaker, my goal is to inspire you to take charge of your life and achieve balance.

Incorporating mindfulness and meditation into your daily routine can be a game-changer, helping you manage stress, gain clarity, and live a more fulfilling life.

Real-Life Example 1:

Jack, a high school teacher, was struggling with balancing his work and personal life. He constantly felt overwhelmed and on the verge of burnout. As an experienced Life Coach, I introduced Jack to mindfulness and meditation practices to help him regain control over his life. Jack started practicing mindful breathing for 10 minutes each day before work and found that his stress levels significantly decreased. He learned to be present in the moment, both in his professional and personal life, leading to improved relationships and a healthier work-life balance.

Real-Life Example 2:

Samantha, a single mother of two, was juggling multiple responsibilities and feeling the weight of the world on her shoulders. As an experienced Consultant, I suggested she try incorporating loving-kindness meditation into her daily routine. Samantha dedicated 15 minutes each night to this practice, focusing on cultivating love and compassion for herself and others. Over time, she noticed a shift in her attitude and outlook on life, becoming more patient, understanding, and emotionally resilient in the face of adversity.

Real-Life Example 3:

Carlos, a college student, was experiencing severe test anxiety, leading to poor academic performance. As a motivational speaker, I shared my own experience with mindfulness and meditation and encouraged Carlos to give it a try. He began practicing a guided meditation app every day for 20 minutes, focusing on deep relaxation and visualization techniques. Within a few weeks, Carlos found that his anxiety had significantly reduced, and he was able to approach his exams with a newfound sense of confidence and calm.

These real-life examples demonstrate the transformative power of mindfulness and meditation. By committing to a consistent practice, individuals from all walks of life can experience improved mental and emotional well-being, leading to a more balanced and fulfilling life. Remember, as an experienced Life Coach, my goal is to guide and empower you on your journey towards self-improvement and personal success.

Q: How can mindfulness and meditation help reduce stress?

A: Mindfulness and meditation can help reduce stress by promoting relaxation, enhancing self-awareness, and cultivating a non-judgmental attitude towards thoughts and emotions.

Q: What is the best time of day to practice meditation?

A: The best time to practice meditation varies for each individual. Some people prefer mornings to start the day with clarity, while others may find evenings more beneficial to unwind and reflect.

Q: How long should I meditate for optimal results?

A: While there is no one-size-fits-all answer, most experts suggest starting with 10-15 minutes per day and gradually increasing the duration as you become more comfortable with the practice.

Q: Can I practice mindfulness while doing daily activities?

A: Yes, mindfulness can be practiced during daily activities by bringing full attention to the present moment and engaging in the task at hand without judgment.

Q: What are some common obstacles to maintaining a consistent meditation practice?

A: Common obstacles include lack of time, difficulty in quieting the mind, and unrealistic expectations. Addressing these challenges by setting aside dedicated time and cultivating patience can help establish a consistent practice.

Q: How can I choose the right meditation technique for me?

A: Experiment with different techniques, such as focused attention, loving-kindness, or body scan meditation, to determine which resonates best with your personal preferences and goals.

Q: Can mindfulness and meditation help with anxiety and depression?

A: Yes, research has shown that mindfulness and meditation can help reduce symptoms of anxiety and depression by fostering emotional regulation, increasing self-awareness, and enhancing positive thinking patterns.

Q: Is it normal for my mind to wander during meditation?

A: Yes, it is normal for the mind to wander. The key is to gently bring your focus back to your chosen object of attention without judgment or frustration.

Q: Are there any apps or resources that can help me get started with mindfulness and meditation?

A: There are numerous apps and resources available, such as Headspace, Calm, and Insight Timer, which offer guided meditations, instructional videos, and mindfulness exercises.

Q: How can I incorporate mindfulness into my busy schedule?

A: Consider setting aside dedicated time for meditation or practicing mindfulness during daily activities, such as eating, walking, or commuting.

Q: What are the long-term benefits of practicing mindfulness and meditation?

A: Long-term benefits include improved mental and emotional well-being, enhanced focus and concentration, better stress management, and increased self-awareness.

Q: Can I practice meditation if I have physical limitations or pain?

A: Yes, there are many meditation techniques that can be adapted to accommodate physical limitations, such as seated or lying down meditation, or using props for support.

Q: How can I stay motivated to maintain my meditation practice?

A: Set realistic expectations, track your progress, connect with a meditation community, and remind yourself of the benefits you experience through your practice.

Q: Can children and teenagers benefit from mindfulness and meditation?

A: Yes, mindfulness and meditation can help children and teenagers develop emotional regulation, enhance focus, and improve overall well-being.

Q: How can I measure my progress with mindfulness and meditation?

A: Progress can be measured by observing changes in your emotional and mental well-being, ability to focus, and overall sense of inner peace.

Q: Is it necessary to sit in a specific posture during meditation?

A: While certain postures are recommended for meditation, the most important aspect is to find a comfortable position that allows for relaxation and alertness.

Q: How can I handle negative emotions or thoughts that arise during meditation?

A: When negative emotions or thoughts arise, acknowledge them without judgment, and gently bring your focus back to your chosen object of attention. Over time, you will develop the ability to observe these thoughts without becoming overwhelmed by them.

Q: Can I combine mindfulness and meditation with other relaxation techniques, such as yoga or deep breathing exercises?

A: Yes, combining mindfulness and meditation with other relaxation techniques can enhance the benefits of both practices, creating a more comprehensive approach to stress management and well-being.

Q: Is it necessary to have a designated meditation space in my home?

A: While having a designated meditation space can be helpful, it is not necessary. The key is to find a quiet and comfortable location where you can practice regularly without distractions.

Q: How can I deal with skepticism from friends or family members regarding the benefits of mindfulness and meditation?

A: Address skepticism by sharing your personal experiences and the research supporting the benefits of mindfulness and meditation. Encourage them to explore the practice themselves, while respecting their opinions and choices.

Technique 1: Mindful Breathing Exercise

Follow these steps for a simple mindful breathing exercise:

Find a quiet, comfortable space where you can sit or lie down.

Close your eyes and bring your awareness to your breath.

Breathe in deeply through your nose, filling your lungs completely.

Exhale slowly through your mouth, releasing all the air from your lungs.

As you breathe, focus on the sensation of the air entering and leaving your body.

When your mind wanders, gently bring it back to your breath.

Practice this exercise for 5-10 minutes daily to cultivate mindfulness and reduce stress.

Technique 2: Body Scan Meditation

Follow these steps for a body scan meditation:

Lie down on your back in a comfortable position.

Close your eyes and take a few deep breaths to center yourself.

Begin at the top of your head and slowly bring your attention to each part of your body, moving downward.

As you focus on each body part, notice any sensations, such as tension or relaxation, without judgment.

Continue scanning your entire body, all the way down to your toes.

Practice this technique for 10-20 minutes to cultivate mindfulness and release tension.

Technique 3: Loving-Kindness Meditation

Follow these steps for a loving-kindness meditation:

Sit comfortably with your eyes closed.

Begin by silently repeating positive phrases towards yourself, such as "May I be happy, may I be healthy, may I be safe, may I be at ease."

After a few minutes, shift your focus to someone you care about, repeating the phrases for them.

Gradually expand your circle of loving-kindness to include acquaintances, strangers, and even those you may have difficulty with.

End the meditation by wishing well for all beings.

Practice this meditation for 10-20 minutes to cultivate compassion and empathy.

Technique 4: Mindfulness Journaling

Follow these steps for mindfulness journaling:

Choose a quiet, comfortable space where you can write without distractions.

Set a timer for 10-15 minutes.

Begin writing about your thoughts, feelings, and experiences, focusing on the present moment.

Write without judgment, allowing yourself to explore your thoughts and emotions fully.

When the timer goes off, read over what you have written, reflecting on any insights or patterns you notice.

Practice this technique daily to cultivate self-awareness and emotional intelligence.

By incorporating these techniques into your daily routine, you will be well on your way to cultivating mindfulness and reaping the benefits of a more balanced life.

In today's fast-paced world, it can be challenging to find the perfect balance between work and personal life. Establishing work-life harmony is crucial for maintaining good mental health, strong relationships, and overall well-being. As a life coach and motivational speaker, I have seen firsthand the positive impact that work-life harmony can have on individuals and their families.

To achieve work-life harmony, consider implementing the following strategies:

Set boundaries: Clearly define your work hours and personal time. This might involve setting a specific time to start and finish work, avoiding work-related tasks outside of these hours, and communicating your boundaries to coworkers and family members.

Prioritize self-care: Take care of your physical, emotional, and mental well-being. This includes exercising regularly, eating a balanced diet, getting enough sleep, and engaging in activities that bring you joy and relaxation.

Manage your time effectively: Use time management techniques, such as creating to-do lists, setting goals, and prioritizing tasks, to maximize productivity during work hours. This will allow you to fully engage in your personal life when you are not working.

Learn to delegate: Recognize that you cannot do everything yourself. Delegate tasks to others when possible, both at work and in your personal life. This will help reduce stress and allow you to focus on the most important tasks.

Develop a support network: Surround yourself with people who understand the importance of work-life harmony and can offer encouragement and assistance when needed. This might include friends, family, or colleagues who share similar goals and values.

Real-life example: Susan, a marketing executive, was struggling to find time for her family and personal interests due to long work hours and constant demands from her job. She decided to establish boundaries by setting specific work hours, delegating tasks to her team, and prioritizing self-care. With the help of her support network, Susan was able to create work-life harmony and enjoy a more fulfilling personal life.

Suggested Readings:

"The One Thing" by Gary Keller and Jay Papasan

"Essentialism: The Disciplined Pursuit of Less" by Greg McKeown

"The 7 Habits of Highly Effective People" by Stephen R. Covey

Questions and answers:

Q: How can I set boundaries between work and personal life when working from home?

A: Establish a dedicated workspace, set specific work hours, and communicate your boundaries with your family and coworkers. It's also essential to take breaks and separate yourself from your work environment during non-work hours.

Q: What if my employer expects me to be available 24/7?

A: It's essential to have an open conversation with your employer about the importance of work-life harmony and how it can benefit both you and the company. Share your concerns and propose a more balanced schedule that allows for personal time and self-care.

Q: How can I prioritize self-care when I have a demanding job and a busy personal life?

A: Schedule self-care activities into your daily routine, just as you would with work tasks. This could include exercise, meditation, hobbies, or spending time with loved ones. Remember that taking care of yourself will make you more productive and better equipped to handle challenges in both your work and personal life.

By implementing these strategies and staying committed to work-life harmony, you'll be on your way to a more balanced and fulfilling life. Remember, it's not about achieving perfection but rather finding the right balance that allows you to thrive in all aspects of your life.

Establishing Work-Life Harmony

Real-Life Story 1:

Dave, a successful entrepreneur, was struggling to find balance in his life. His business was growing rapidly, but it was taking a toll on his personal life. He barely spent time with his family, and his health was starting to decline.

After attending a motivational seminar, Dave decided to take control of his life. He sought guidance from an experienced life coach, who helped him prioritize his goals and set boundaries between work and personal life. Dave delegated some of his responsibilities to his team and began scheduling quality time with his family. As a result, he was able to establish work-life harmony, and his relationships and health improved significantly.

Real-Life Story 2:

Maria, a dedicated teacher, was constantly overwhelmed with her workload. She would often bring work home, leaving her with little time for her husband and two children. Maria felt guilty and stressed, which affected her mental well-being and family dynamics.

Maria consulted with an experienced consultant who specialized in work-life harmony. Together, they devised a plan that included setting boundaries, managing her time effectively, and practicing self-care. Maria started attending a yoga class twice a week and made it a point to have dinner with her family every night. As she implemented these changes, Maria noticed a significant improvement in her overall well-being and the quality of her relationships with her family.

Real-Life Story 3:

Chris, a high-performing sales manager, was always on the road, traveling for work. His wife and children missed him, and he felt disconnected from his loved ones. The stress of trying to balance work and family life was taking a toll on his happiness.

Chris reached out to a motivational speaker who specialized in work-life harmony. Through their conversations, Chris realized the importance of quality time with his family and maintaining a healthy lifestyle. He negotiated a more flexible work schedule with his employer and committed to regular family vacations. By incorporating these changes, Chris was able to achieve work-life harmony and strengthen his bond with his family.

These real-life stories illustrate the power of prioritizing work-life harmony and seeking guidance from experienced professionals. By implementing the strategies and techniques shared in this chapter, you too can achieve balance and enjoy a more fulfilling personal and professional life.

Q: What is the first step towards achieving work-life harmony?

A: The first step is to recognize the need for balance and prioritize your personal and professional goals.

Q: How can I set boundaries between work and personal life?

A: Learn to say no, delegate tasks, create a schedule, and communicate your needs to your employer and family.

Q: How can I improve my time management skills?

A: Use tools such as calendars, to-do lists, and time-tracking apps. Also, practice prioritization and break tasks into smaller, manageable chunks.

Q: How can I practice self-care while managing a busy schedule?

A: Schedule regular breaks, exercise, eat well, and prioritize sleep. Also, make time for hobbies and interests outside of work.

Q: What role does communication play in work-life harmony?

A: Open and honest communication with your employer, family, and friends helps manage expectations and creates understanding and support.

Q: How can I maintain a healthy work-life harmony when working from home?

A: Establish a designated workspace, set work hours, take breaks, and maintain communication with colleagues and family members.

Q: Can flexible work arrangements contribute to work-life harmony?

A: Yes, flexible work arrangements allow you to adjust your work schedule to better accommodate personal and family needs.

Q: How do I know if I need professional help to achieve work-life harmony?

A: If you're struggling with stress, anxiety, or burnout, and self-help strategies aren't working, consider seeking help from a life coach or therapist.

Q: How can I ensure work-life harmony while traveling for work?

A: Plan ahead, maintain communication with loved ones, and schedule personal time during your trips.

Q: How can I involve my family in the process of establishing work-life harmony?

A: Have open discussions about your goals, expectations, and needs. Encourage your family members to share their thoughts and feelings as well.

Q: Can technology help or hinder work-life harmony?

A: Technology can help by improving organization and communication but can also hinder if it leads to constant availability and work-related distractions.

Q: How can I measure my progress towards achieving work-life harmony?

A: Assess your stress levels, overall happiness, quality of relationships, and personal satisfaction regularly.

Q: Can a healthy work-life harmony improve job performance?

A: Yes, when employees have a balanced life, they tend to be more focused, engaged, and productive at work.

Q: What is the role of my employer in establishing work-life harmony?

A: Employers can support work-life harmony by offering flexible work arrangements, promoting a healthy work culture, and providing resources for employee well-being.

Q: How can I manage work-life harmony during periods of high stress or workload?

A: Prioritize tasks, delegate responsibilities, maintain self-care practices, and communicate your needs with your support network.

Q: How do I know if I have achieved work-life harmony?

A: You will likely experience lower stress levels, increased satisfaction in both personal and professional life, and healthier relationships.

Q: How can I maintain work-life harmony in the long term?

A: Regularly reassess your priorities, set boundaries, practice self-care, and adapt to life changes.

Q: Can hobbies and leisure activities contribute to work-life harmony?

A: Yes, engaging in hobbies and leisure activities can reduce stress, improve mental well-being, and provide a sense of balance.

Q: How can I support my colleagues in achieving work-life harmony?

A: Encourage open communication, be understanding and empathetic, offer assistance when needed, and promote a positive work culture.

Q: What is the role of personal values in establishing work-life harmony?

A: Identifying and prioritizing your personal values helps you make decisions that align with your goals, ultimately contributing to a sense of harmony between work and personal life.

Technique 1: The 80/20 Rule

One effective approach to establishing work-life harmony is to apply the 80/20 rule, also known as the Pareto Principle.

This rule states that 80% of your results come from 20% of your efforts. Identify the most crucial tasks in your work and personal life and focus on those. By doing so, you can achieve a greater sense of balance and fulfillment.

Technique 2: Time Blocking

Time blocking involves scheduling dedicated blocks of time for specific tasks, both professional and personal. By assigning specific time slots for work, family, hobbies, and self-care, you can create a clear boundary between different aspects of your life, leading to improved work-life harmony.

Technique 3: Digital Detox

Disconnecting from technology at specific times during the day or week can help reduce stress and improve your overall well-being. Set boundaries for your screen time, especially during non-working hours, to cultivate a sense of presence and mindfulness in your personal life.

Technique 4: The Eisenhower Matrix

The Eisenhower Matrix is a time management tool that helps you prioritize tasks based on their importance and urgency. By categorizing tasks into four quadrants (urgent and important, important but not urgent, urgent but not important, and neither urgent nor important), you can allocate your time and energy more effectively, resulting in better work-life harmony.

Technique 5: Gratitude Journaling

Maintaining a gratitude journal can help you focus on the positive aspects of your life, leading to a greater sense of fulfillment and balance. Set aside time each day to write down three things you're grateful for, both in your professional and personal life. This practice can help shift your mindset and contribute to overall work-life harmony.

In conclusion, by incorporating these practical techniques into your daily routine, you can make significant strides toward establishing work-life harmony. Remember to be patient with yourself as you implement these changes, as finding balance is a continuous process that requires time and effort. Stay committed to your goals and remember that achieving harmony between your work and personal life is crucial for overall well-being and success.

Prioritizing Self-Care and Self-Compassion

In this section, we will delve into the importance of prioritizing self-care and self-compassion as a means to achieve balance in life. As a life coach and motivational speaker, I have observed that individuals who make self-care and self-compassion a priority are more successful in managing stress, maintaining healthy relationships, and achieving their goals.

Self-care involves taking deliberate actions to maintain and improve one's physical, emotional, and mental well-being. It's important to understand that self-care is not selfish or indulgent; rather, it is a necessary component of a balanced life. Some examples of self-care activities include regular exercise, proper nutrition, adequate sleep, and engaging in hobbies that bring joy and relaxation.

Self-compassion, on the other hand, involves treating oneself with kindness and understanding, especially during difficult times. It is the practice of extending the same level of empathy and care to oneself that we would offer to a close friend or loved one. By cultivating self-compassion, we can better navigate life's challenges and develop greater resilience.

Here are some tips and strategies to prioritize self-care and self-compassion in your daily life:

Schedule self-care activities: Just as you would schedule a meeting or an appointment, pencil in time for self-care activities. This can be as simple as a daily walk, a weekly yoga class, or a monthly spa day. Prioritizing self-care helps ensure that it becomes a regular part of your routine.

Set boundaries: Establishing boundaries with work, family, and friends is crucial for maintaining balance in life. Communicate your needs and limits clearly, and learn to say "no" when necessary to protect your well-being.

Develop a self-compassion mantra: Create a personal mantra that you can repeat to yourself during times of stress or self-doubt. This could be something like, "I am doing my best, and that is enough," or "I am deserving of love and care, just like everyone else."

Practice mindfulness: Mindfulness techniques, such as meditation and deep breathing exercises, can help cultivate self-awareness and self-compassion. Set aside time each day to practice mindfulness and observe your thoughts and emotions without judgment.

Seek support: Reach out to friends, family, or a professional therapist when you need help or encouragement. Building a support network can make it easier to prioritize self-care and self-compassion.

Suggested readings:

"The Mindful Self-Compassion Workbook" by Kristin Neff and Christopher Germer

"Self-Compassion: The Proven Power of Being Kind to Yourself" by Kristin Neff

"The Self-Care Solution" by Jennifer Ashton

Real-life example:

Emma, a busy marketing executive, was struggling with burnout and feeling overwhelmed by her personal and professional responsibilities. She reached out to a life coach for guidance. Through coaching sessions, Emma learned the importance of self-care and self-compassion. She began to schedule regular self-care activities, such as yoga classes and monthly massages, and practiced setting boundaries with her colleagues and family members. By cultivating self-compassion, Emma found that she was better able to manage stress, improve her relationships, and achieve a greater sense of balance in her life.

Real-life story 1:

James, a devoted father and successful entrepreneur, always put his family and business first. Unfortunately, this left little time for self-care or self-compassion. He came to me as a life coach, seeking help in managing his overwhelming schedule.

After discussing James' situation, we identified the areas where he could incorporate self-care into his daily routine. He began waking up an hour earlier to exercise and meditate, started eating healthier meals, and set aside time each week for a relaxing activity he enjoyed. James also learned to practice self-compassion by reminding himself that he was doing the best he could for his family and business, and that it was okay to prioritize his own well-being.

As a result, James noticed improvements in his energy levels, overall mood, and relationships with his family and employees. By prioritizing self-care and self-compassion, he was able to achieve a greater sense of balance in his life.

Real-life story 2:

Sophia, a dedicated nurse, was constantly sacrificing her own needs to care for others. Her selfless nature and demanding job left her feeling exhausted and burnt out. She contacted me as a motivational speaker, looking for guidance on how to find balance in her life.

Together, we explored ways Sophia could prioritize her own well-being while still meeting her professional obligations. She began taking short breaks throughout her shifts to practice deep breathing exercises, setting boundaries with colleagues, and carving out time on her days off to engage in hobbies that brought her joy. We also worked on cultivating self-compassion by helping Sophia recognize the importance of her role as a nurse and understanding that she deserved care and compassion too.

As she began implementing these changes, Sophia experienced a significant decrease in her stress levels, improved sleep, and a renewed passion for her career. By prioritizing self-care and self-compassion, she found the balance she was seeking and became an even more effective caregiver.

Real-life story 3:

Michael, a high-achieving law student, struggled with self-criticism and an overwhelming need for perfection. He reached out to me as a life coach, seeking help in managing his stress and finding balance in his life.

We worked together to develop a self-compassion mantra that Michael could repeat to himself during moments of self-doubt: "I am worthy of love and understanding, even when I make mistakes." He also began to incorporate self-care activities into his daily routine, such as journaling, spending time in nature, and practicing mindfulness meditation.

Over time, Michael noticed that his anxiety and stress levels decreased, his academic performance improved, and his relationships with his peers and family members became more enjoyable. By embracing self-care and self-compassion, Michael achieved the balance he desired and became better equipped to handle the challenges of law school and beyond.

Q: How do I recognize when I need to prioritize self-care and self-compassion?

A: Signs that you need to prioritize self-care and self-compassion include feeling constantly exhausted, increased stress levels, difficulty concentrating, irritability, and a decline in overall well-being. It's essential to pay attention to these signals and take steps to care for yourself.

Q: What are some simple self-care practices I can incorporate into my daily routine?

A: Simple self-care practices include getting enough sleep, eating well, exercising, staying hydrated, practicing mindfulness meditation, and engaging in hobbies or activities you enjoy.

Q: How do I practice self-compassion when I feel overwhelmed by self-criticism?

A: Self-compassion involves recognizing your own worth, accepting your imperfections, and being kind to yourself during difficult moments. You can practice self-compassion by using positive affirmations, journaling about your feelings, or seeking support from friends or a therapist.

Q: How can I create a self-care routine that works for my schedule and lifestyle?

A: To create a self-care routine, start by identifying activities that help you feel relaxed and rejuvenated. Then, set aside specific times during your day or week to engage in these activities, making them a non-negotiable part of your schedule.

Q: Can self-care and self-compassion improve my relationships with others?

A: Yes, prioritizing self-care and self-compassion can lead to improved emotional well-being, which can positively impact your relationships with others. When you feel better about yourself, you're better equipped to engage with others in a healthy and supportive manner.

Q: How can I set boundaries to protect my self-care time?

A: Setting boundaries involves communicating your needs clearly and assertively, saying "no" when necessary, and respecting your own limits. It's essential to prioritize your well-being and make self-care a non-negotiable aspect of your life.

Q: Why is it important to prioritize self-care even when I'm busy?

A: Prioritizing self-care during busy times can prevent burnout, improve your overall well-being, and increase your productivity. Taking care of yourself helps ensure you have the energy and mental clarity to tackle your tasks effectively.

Q: What are some ways to practice self-compassion during challenging times?

A: During challenging times, practice self-compassion by acknowledging your emotions, offering yourself kindness and understanding, and remembering that everyone experiences difficulties. You can also engage in activities that promote relaxation and emotional healing.

Q: Can practicing self-compassion improve my mental health?

A: Yes, self-compassion has been linked to improved mental health outcomes, including reduced anxiety, depression, and stress. By offering yourself kindness and understanding, you create a supportive environment for personal growth and emotional well-being.

Q: How can I remind myself to practice self-care and self-compassion regularly?

A: To remind yourself to practice self-care and self-compassion, set alarms or calendar reminders, create visual cues (such as sticky notes), enlist the support of friends or family, and make self-care activities a regular part of your routine.

Q: What role does mindfulness play in self-care and self-compassion?

A: Mindfulness is the practice of being present and fully engaged in the current moment. It can help you become more aware of your thoughts and feelings, allowing you to respond with greater self-compassion and intentionality when engaging in self-care activities.

Q: How do I recognize when I need to prioritize self-care and self-compassion?

A: Signs that you need to prioritize self-care and self-compassion include feeling constantly exhausted, increased stress levels, difficulty concentrating, irritability, and a decline in overall well-being. It's essential to pay attention to these signals and take steps to care for yourself.

Q: What are some simple self-care practices I can incorporate into my daily routine?

A: Simple self-care practices include getting enough sleep, eating well, exercising, staying hydrated, practicing mindfulness meditation, and engaging in hobbies or activities you enjoy.

Q: How do I practice self-compassion when I feel overwhelmed by self-criticism?

A: Self-compassion involves recognizing your own worth, accepting your imperfections, and being kind to yourself during difficult moments. You can practice self-compassion by using positive affirmations, journaling about your feelings, or seeking support from friends or a therapist.

Q: How can I create a self-care routine that works for my schedule and lifestyle?

A: To create a self-care routine, start by identifying activities that help you feel relaxed and rejuvenated. Then, set aside specific times during your day or week to engage in these activities, making them a non-negotiable part of your schedule.

Q: Can self-care and self-compassion improve my relationships with others?

A: Yes, prioritizing self-care and self-compassion can lead to improved emotional well-being, which can positively impact your relationships with others. When you feel better about yourself, you're better equipped to engage with others in a healthy and supportive manner.

Q: How can I set boundaries to protect my self-care time?

A: Setting boundaries involves communicating your needs clearly and assertively, saying "no" when necessary, and respecting your own limits. It's essential to prioritize your well-being and make self-care a non-negotiable aspect of your life.

Q: Why is it important to prioritize self-care even when I'm busy?

A: Prioritizing self-care during busy times can prevent burnout, improve your overall well-being, and increase your productivity. Taking care of yourself helps ensure you have the energy and mental clarity to tackle your tasks effectively.

Q: What are some ways to practice self-compassion during challenging times?

A: During challenging times, practice self-compassion by acknowledging your emotions, offering yourself kindness and understanding, and remembering that everyone experiences difficulties. You can also engage in activities that promote relaxation and emotional healing.

Q: Can practicing self-compassion improve my mental health?

A: Yes, self-compassion has been linked to improved mental health outcomes, including reduced anxiety, depression, and stress. By offering yourself kindness and understanding, you create a supportive environment for personal growth and emotional well-being.

Q: How can I remind myself to practice self-care and self-compassion regularly?

A: To remind yourself to practice self-care and self-compassion, set alarms or calendar reminders, create visual cues (such as sticky notes), enlist the support of friends or family, and make self-care activities a regular part of your routine.

Q: What role does mindfulness play in self-care and self-compassion?

A: Mindfulness is the practice of being present and fully engaged in the current moment. It can help you become more aware of your thoughts and feelings, allowing you to respond with greater self-compassion and intentionality when engaging in self-care activities.

Q: How can I cultivate self-compassion if I have a history of negative self-talk?

A: To cultivate self-compassion when you have a history of negative self-talk, begin by becoming more aware of your inner critic and challenging negative thoughts with kindness and understanding. Practice replacing harsh self-judgment with positive affirmations and engage in activities that promote self-love and acceptance.

Q: Can self-care and self-compassion improve my work performance?

A: Yes, prioritizing self-care and self-compassion can lead to increased focus, productivity, and overall work performance. When you take care of yourself mentally and physically, you're better equipped to handle the demands of your job.

Q: How can I encourage a self-care and self-compassion mindset in my workplace?

A: Encourage a self-care and self-compassion mindset at work by modeling healthy behavior, fostering open communication, providing resources for stress management, and advocating for work-life balance policies.

Q: How do I know if I'm practicing enough self-care and self-compassion?

A: Signs that you're practicing enough self-care and self-compassion include feeling more relaxed, reduced stress levels, increased focus and productivity, and a general sense of well-being. If you notice improvements in these areas, it's likely that you're taking sufficient care of yourself.

Q: What are some strategies for incorporating self-care and self-compassion into a busy schedule?

A: Strategies for incorporating self-care and self-compassion into a busy schedule include setting specific times for self-care activities, breaking tasks into smaller, more manageable pieces, delegating or outsourcing when possible, and reminding yourself of the importance of taking care of your well-being.

Q: Can I practice self-compassion even if I feel like I don't deserve it?

A: Yes, you can and should practice self-compassion even if you feel like you don't deserve it. Remember that self-compassion is not about whether you're worthy, but rather about treating yourself with kindness and understanding during difficult times.

Q: How can I teach my children the importance of self-care and self-compassion?

A: Teach your children the importance of self-care and self-compassion by modeling healthy behaviors, engaging in open conversations about emotions and well-being, and encouraging them to participate in self-care activities that promote relaxation and self-love.

Q: What are some self-compassion exercises I can try when I'm feeling overwhelmed?

A: When feeling overwhelmed, try self-compassion exercises such as writing a compassionate letter to yourself, practicing loving-kindness meditation, or engaging in soothing self-care activities like a warm bath or a walk in nature.

Q: How can I maintain my self-care and self-compassion practices long-term?

A: To maintain your self-care and self-compassion practices long-term, make them a consistent part of your daily routine, be flexible and open to adjusting your practices as needed, and seek support from friends, family, or professionals when necessary.

As an experienced life coach and motivational speaker, I understand the importance of prioritizing self-care and self-compassion in our lives. In this section, I will provide you with practical tools, exercises, and techniques that will help you put our advice into practice and achieve balance in your life.

Self-Care Audit: Assess your current self-care habits by tracking your daily activities for a week. Note the activities that contribute to your well-being and those that deplete your energy. Reflect on your findings and identify areas for improvement.

Personal Self-Care Plan: Develop a self-care plan that outlines specific activities and goals in different areas of your life, such as physical, emotional, social, and spiritual self-care. Revisit and update your plan regularly.

Mindful Journaling: Engage in daily journaling to explore your emotions, thoughts, and feelings. This practice promotes self-awareness and helps you identify patterns of negative self-talk.

Gratitude Practice: Every day, list three things you're grateful for. This practice fosters a positive mindset and increases self-compassion.

Loving-Kindness Meditation: Spend 10-15 minutes each day practicing loving-kindness meditation. Focus on cultivating feelings of love and compassion for yourself and others.

Establish Boundaries: Learn to set healthy boundaries in your personal and professional life, and communicate them clearly to others.

Affirmations: Create a list of positive affirmations to repeat daily, reinforcing your self-worth and promoting self-compassion.

Time Management Techniques: Implement effective time management strategies, such as the Pomodoro Technique or the Eisenhower Matrix, to ensure you have time for self-care activities.

Digital Detox: Schedule regular breaks from technology to reduce stress and increase mindfulness.

Mindful Movement: Engage in physical activities that promote mindfulness, such as yoga, tai chi, or gentle stretching.

Emotional Self-Care Toolkit: Create a list of activities that help you manage your emotions, such as deep breathing exercises, visualization, or engaging in creative hobbies.

Social Support System: Cultivate a strong support system by connecting with friends, family, or support groups who understand the importance of self-care and self-compassion.

Sleep Hygiene: Prioritize quality sleep by establishing a consistent sleep schedule and creating a relaxing bedtime routine.

Nutrition: Pay attention to your diet and nourish your body with balanced, nutrient-rich meals.

Acts of Kindness: Regularly engage in acts of kindness towards yourself and others, fostering a compassionate mindset.

Guided Imagery: Practice guided imagery exercises to promote relaxation and self-compassion.

Self-Compassion Break: When faced with a challenging situation, take a brief self-compassion break to remind yourself of your worth and validate your feelings.

Creative Expression: Engage in creative activities, such as painting, writing, or dancing, to express your emotions and practice self-care.

Nature Therapy: Spend time in nature to reduce stress, increase mindfulness, and foster self-compassion.

Professional Help: Seek professional support from a therapist, life coach, or counselor if you're struggling to prioritize self-care and self-compassion in your life.

As an experienced life coach, consultant, and motivational speaker, I know the value of prioritizing self-care and self-compassion. To help you achieve balance in your life, I have crafted a collection of practical tools, exercises, and techniques that you can follow to put the advice into practice. Remember, this is an instruction for you to read and follow directions on your own. Here, we'll introduce worksheets, templates, and guided meditations to support your journey.

Self-Care Assessment Worksheet: Use this worksheet to evaluate your current self-care habits. Identify areas in which you excel and areas that need improvement. This exercise will help you become more mindful of your daily routine.

Personalized Self-Care Blueprint: Create a customized self-care plan that caters to your unique needs and preferences. Develop a schedule that includes physical, emotional, social, and spiritual self-care activities.

Mindful Breathing Exercises: Incorporate mindfulness into your daily routine with simple breathing exercises. Practice deep, intentional breaths to help you relax and focus.

Self-Compassion Letter Writing: Write a heartfelt letter to yourself, acknowledging your strengths, accomplishments, and areas for growth. This exercise promotes self-compassion and self-acceptance.

Guided Meditation Scripts: Follow along with guided meditation scripts to cultivate a sense of inner peace and self-awareness. Use these resources to develop a daily meditation practice.

Boundaries Template: Use a template to help you establish healthy boundaries in your personal and professional life.

Clearly define your limits and communicate them effectively to others.

Positive Affirmation Cards: Create a set of positive affirmation cards that resonate with you. Keep these cards handy and review them regularly to reinforce your self-worth and encourage self-compassion.

Time Management Planner: Develop a time management planner to organize your schedule effectively, ensuring you have time for self-care activities.

Digital Detox Plan: Design a plan to reduce your screen time and engage in more mindful activities, such as reading or spending time in nature.

Movement and Mindfulness Log: Track your physical activities that promote mindfulness, like yoga, tai chi, or stretching. Use this log to monitor your progress and make adjustments as needed.

Emotional Self-Care Toolkit: Create a toolkit of activities and resources that help you manage your emotions effectively. This toolkit may include breathing exercises, visualization techniques, or creative hobbies.

Social Support Network List: Maintain a list of supportive friends, family members, and professional contacts. Reach out to these individuals when you need encouragement or guidance.

Sleep Hygiene Checklist: Develop a sleep hygiene checklist that outlines your bedtime routine and habits that promote restful sleep.

Nutritional Meal Planner: Create a meal planner that includes balanced, nutrient-rich meals to nourish your body and mind.

Kindness Challenge Calendar: Design a calendar that encourages daily acts of kindness towards yourself and others, fostering a compassionate mindset.

Relaxation Techniques Guide: Compile a list of relaxation techniques, such as progressive muscle relaxation, guided imagery, or deep breathing exercises.

Self-Compassion Journal: Keep a journal dedicated to documenting your self-compassion journey, including your thoughts, feelings, and progress.

Creative Expression Journal: Use a separate journal to express your emotions through creative means, such as painting, writing, or dancing.

Nature Therapy Plan: Plan regular outings in nature to reduce stress, increase mindfulness, and practice self-compassion.

Professional Help Directory: Keep a list of therapists, life coaches, or counselors who specialize in self-care and self-compassion. Reach out to these professionals if you need additional support.

Chapter 5: Unstoppable Momentum: Continuously Evolving and Adapting

Embracing Lifelong Learning

As an experienced life coach, motivational speaker, and consultant, I cannot stress enough the importance of embracing lifelong learning. It's crucial for personal growth, professional success, and overall well-being. In this section, we will explore the benefits of continuous learning, strategies to incorporate learning into your daily life, and ways to stay motivated and adapt to change.

One of the keys to unstoppable momentum in life is adopting a mindset of continuous growth and development. Lifelong learning enables you to adapt to new situations, remain relevant in your career, and stay mentally sharp. It allows you to approach challenges with confidence and resilience, ultimately leading to a more fulfilling and successful life.

To make lifelong learning an integral part of your routine, consider the following strategies:

Set clear learning goals: Identify the areas in which you want to grow and develop. Establish specific, measurable, achievable, relevant, and time-bound (SMART) goals to guide your learning journey.

Diversify your learning sources: To gain a well-rounded perspective, explore various learning sources such as books, online courses, workshops, seminars, podcasts, and networking events.

Create a learning schedule: Allocate time in your daily routine for learning activities. This can be as simple as setting aside 30 minutes each day to read, listen to a podcast, or watch a TED Talk.

Join a community of learners: Surround yourself with like-minded individuals who share your passion for personal growth. Engage in group discussions, attend events, and exchange ideas with your peers.

Apply your knowledge: Put your newly acquired skills and knowledge into practice. This not only reinforces your learning but also allows you to assess your progress and adjust your approach as needed.

Reflect on your learning journey: Regularly assess your progress and identify areas for improvement. Celebrate your achievements and use setbacks as opportunities for growth.

As you embark on your lifelong learning journey, you may encounter challenges that test your motivation and dedication. Remember that as a motivational speaker, I've witnessed countless individuals overcome obstacles and reach their full potential. Here are some real-life examples to inspire you:

Example 1:

Problem: Sarah struggled to find the motivation to learn new skills in her free time.

Solution: Sarah discovered her passion for photography and enrolled in an online course. The excitement of learning something she loved helped her stay committed to her learning journey.

Example 2:

Problem: James had difficulty finding time to learn amidst his busy schedule.

Solution: James started listening to educational podcasts during his daily commute, effectively turning downtime into a valuable learning opportunity.

To further support your learning journey, consider these recommended resources:

"Mindset: The New Psychology of Success" by Carol S. Dweck: This book explores the concept of growth mindset and provides practical advice on fostering a love for learning.

Coursera (www.coursera.org): This online platform offers thousands of courses across a wide range of subjects, enabling you to learn at your own pace from top universities and institutions.

TED Talks (www.ted.com): Discover inspiring talks from experts in various fields, covering topics such as technology, science, personal development, and more.

Embracing lifelong learning is a key component of unstoppable momentum. By continuously evolving and adapting, you can achieve greater success, personal fulfillment, and overall well-being. Begin your journey today and unlock your full potential.

Example 1: Overcoming Age Barriers in Lifelong Learning

Problem: Susan, a 55-year-old woman, felt that she was too old to learn new skills and believed that she could not keep up with younger colleagues at work.

Solution: As her life coach, I encouraged Susan to focus on her strengths and view her age as an advantage. We worked together to identify areas where she could enhance her skillset and stay competitive in the workforce. Susan took on the challenge and enrolled in a coding course, proving to herself and her colleagues that age is no barrier to learning and personal growth.

Example 2: Navigating a Career Change

Problem: After being laid off from his job, David struggled to find new employment in his field. He was unsure about his next steps and felt overwhelmed by the prospect of starting over in a new industry.

Solution: David reached out to a career consultant who suggested that he explore his interests and passions to identify a new career path. After some introspection, David realized his love for gardening and horticulture. He pursued a certification in landscape design and, through continuous learning, successfully transitioned into a fulfilling new career.

Example 3: Overcoming Learning Disabilities

Problem: Emily, a young adult with dyslexia, had always struggled academically. She feared that her learning disability would prevent her from furthering her education and career.

Solution: Emily attended a motivational seminar where she was inspired by the speaker's message of perseverance and adaptability. With renewed determination, Emily sought out resources tailored to her learning needs and enrolled in an online graphic design course that utilized visual teaching methods. By embracing lifelong learning and finding strategies that worked for her, Emily was able to overcome her challenges and excel in her chosen field.

Example 4: Balancing Family Life and Personal Development

Problem: As a busy father of three, Mark found it difficult to dedicate time to his personal growth and learning.

Solution: With guidance from an experienced consultant, Mark learned to prioritize his time effectively and create a schedule that allowed for self-improvement without compromising his responsibilities as a father. He began waking up an hour earlier each day to read books or take online courses, demonstrating that it is possible to balance family life with personal development through careful planning and dedication.

These real-life stories illustrate the power of embracing lifelong learning, regardless of one's age, circumstances, or challenges. By continuously evolving and adapting, individuals can overcome obstacles, achieve their goals, and lead fulfilling lives.

Q: How can I maintain the motivation to continuously learn and adapt in a fast-paced world?

A: Set specific, achievable goals for yourself and regularly track your progress. Surround yourself with like-minded individuals who share your passion for learning and stay curious about new developments in your field.

Q: What are some effective ways to incorporate lifelong learning into my daily routine?

A: You can try reading books, listening to podcasts, taking online courses, attending webinars or workshops, and connecting with mentors or experts in your field.

Q: Is it too late for me to start embracing lifelong learning?

A: It's never too late to learn and grow. No matter your age or circumstances, you can always benefit from expanding your knowledge and skills.

Q: How can I overcome my fear of failure when learning new skills?

A: Focus on the learning process rather than the outcome, and practice self-compassion. Remember that everyone makes mistakes and it is a natural part of the learning journey.

Q: How do I choose the right learning resources for my needs?

A: Assess your learning style, identify your goals, and explore various resources to find the ones that resonate with you and help you achieve your objectives.

Q: Can I learn multiple skills simultaneously or should I focus on one skill at a time?

A: It depends on your personal preferences and capacity. Some people thrive when learning multiple skills simultaneously, while others prefer to focus on one skill at a time to avoid feeling overwhelmed.

Q: How can I measure my progress in lifelong learning?

A: Set specific milestones and evaluate your progress regularly. Celebrate your achievements and adjust your goals as needed to ensure continuous growth.

Q: How can I ensure that I am adapting to changes in my field or industry?

A: Stay updated on industry trends and developments by following relevant news, attending conferences, and networking with professionals in your field.

Q: How can I balance my professional and personal life with my commitment to lifelong learning?

A: Prioritize your time effectively, set realistic goals, and create a schedule that allows for both personal and professional growth.

Q: Can I benefit from lifelong learning even if I'm already successful in my career?

A: Absolutely! Lifelong learning can help you stay ahead of industry trends, expand your skillset, and maintain a competitive edge in the job market.

Q: How do I find the motivation to learn when I'm feeling overwhelmed or burnt out?

A: Take a step back and assess your current workload. It's important to practice self-care and ensure that you're not taking on too much. Once you've regained your balance, find learning opportunities that genuinely excite and inspire you.

Q: How can I develop a growth mindset for lifelong learning?

A: Embrace challenges, view setbacks as opportunities for growth, and maintain a positive attitude towards learning and self-improvement.

Q: How do I deal with the fear of being left behind in the rapidly evolving world?

A: By embracing lifelong learning and continuously adapting to change, you can alleviate this fear and stay relevant in your field.

Q: What is the role of networking in lifelong learning?

A: Networking enables you to learn from others' experiences, gain new perspectives, and stay updated on industry trends and opportunities.

Q: How can I maintain my focus and discipline when learning independently?

A: Set clear goals, create a dedicated learning space, and establish a routine that incorporates regular study sessions.

Q: How can I overcome self-doubt and imposter syndrome in my learning journey?

A: Acknowledge your achievements, practice

Q: How can I find a mentor to support my lifelong learning journey?

A: Seek out professionals in your field who inspire you, attend networking events, or join online communities related to your area of interest. Don't hesitate to reach out and ask for guidance or mentorship.

Q: How can I stay motivated when I encounter obstacles in my learning process?

A: Remind yourself of your goals, recognize your achievements thus far, and be patient with yourself. Seek support from friends, family, or mentors, and remember that setbacks are a natural part of the learning journey.

Q: What role does curiosity play in lifelong learning?

A: Curiosity drives you to explore new ideas, expand your knowledge, and develop a deeper understanding of the world around you. By staying curious, you will naturally be more engaged and motivated in your learning endeavors.

Q: How can I ensure that the skills I learn today will remain relevant in the future?

A: Focus on developing transferable skills, such as critical thinking, problem-solving, and adaptability. These skills will help you navigate changes in your field or industry and enable you to continue learning and evolving throughout your career.

As a Life Coach and Motivational Speaker, I want to help you embrace lifelong learning and create unstoppable momentum in your life. Here are some practical tools, exercises, and techniques you can use to put my advice into practice:

Personal Learning Plan: Create a personal learning plan that outlines your goals, the skills you want to develop, and the resources you'll use to achieve them. Regularly review and update this plan to ensure it remains relevant and aligned with your aspirations.

Habit Tracker: Develop a habit tracker to monitor your progress towards your learning goals. Consistency is key, so keep track of your daily efforts to make learning a regular part of your life.

Mind Mapping: Use mind mapping to explore new ideas and connect them to your existing knowledge. This visual tool can help you see patterns, relationships, and gaps in your understanding, stimulating further learning.

Journaling: Start a learning journal to document your progress, insights, and challenges. Reflect on your experiences and use your journal as a source of motivation and inspiration.

Online Communities: Join online forums or social media groups related to your areas of interest. Engage with others, ask questions, and share your own insights to expand your knowledge and network.

Time Management Techniques: Use time management techniques, such as the Pomodoro Technique, to allocate dedicated time to learning without feeling overwhelmed.

Reading List: Create a reading list of books, articles, and other resources relevant to your learning goals. Dedicate time each day to read and absorb new information.

Skill Swap: Connect with others who have complementary skills and arrange skill-swap sessions, where you teach each other something new. This way, you both benefit from each other's expertise.

Podcasts and Webinars: Subscribe to podcasts and attend webinars related to your areas of interest. These can be a great source of inspiration and knowledge.

Reflective Practice: Regularly engage in reflective practice to evaluate your learning progress, identify areas for improvement, and celebrate your achievements. This introspection will help you stay focused and motivated on your lifelong learning journey.

By following these techniques and tools, you'll be well on your way to embracing lifelong learning and achieving unstoppable momentum in your personal and professional life. Remember to be patient with yourself, enjoy the process, and keep pushing forward.

CELEBRATING PROGRESS AND OVERCOMING SETBACKS

As a Life Coach, I understand that the journey towards personal growth and success is rarely linear. Along the way, you will encounter both progress and setbacks. To maintain unstoppable momentum, it is essential to celebrate your progress and learn how to overcome setbacks effectively. In this section, we will discuss the importance of recognizing achievements, dealing with setbacks, and adopting a growth mindset that fosters continuous improvement.

The Power of Celebrating Progress:

Acknowledging and celebrating your progress is crucial for maintaining motivation and cultivating a positive mindset. When you recognize your achievements, no matter how small, you build self-confidence and reinforce your belief in your abilities. Celebrate your progress by setting milestones, rewarding yourself, or sharing your accomplishments with others. This positive reinforcement will encourage you to keep striving towards your goals.

Suggested Reading: "The Progress Principle: Using Small Wins to Ignite Joy, Engagement, and Creativity at Work" by Teresa Amabile and Steven Kramer.

Overcoming Setbacks:

Setbacks are inevitable on the path to success, but it's how you handle them that defines your journey. When faced with setbacks, it's essential to take a step back and assess the situation. Identify the cause of the setback and determine what you can learn from it. Use this knowledge to adapt your approach and overcome obstacles more effectively in the future.

Real-life example: After losing her job, Sarah used the setback as an opportunity to reassess her career goals and pursue her passion for entrepreneurship. She learned from her previous experiences and applied that knowledge to her new venture, ultimately finding success and personal fulfillment.

Adopting a Growth Mindset:

A growth mindset is the belief that your abilities can be developed through dedication and hard work. Embracing this mindset is crucial for overcoming setbacks and continuously evolving. Focus on learning from your experiences, both positive and negative, and view challenges as opportunities for growth.

Question: How can you cultivate a growth mindset in your daily life?

Answer: By embracing challenges, learning from criticism, and focusing on the process rather than the outcome, you can develop a growth mindset that fosters continuous improvement.

Building Resilience:

Resilience is the ability to bounce back from setbacks and adapt to change. Cultivate resilience by reframing negative experiences, seeking support, and practicing self-compassion. Remember that setbacks are temporary, and you have the power to overcome them and continue on your journey.

Suggested Reading: "Option B: Facing Adversity, Building Resilience, and Finding Joy" by Sheryl Sandberg and Adam Grant.

Staying Accountable:

Maintain your momentum by holding yourself accountable for your progress and setbacks. Share your goals with a trusted friend, family member, or mentor who can provide support and encouragement. Regularly review your progress, celebrate your achievements, and adjust your approach as needed.

By celebrating your progress, overcoming setbacks, and adopting a growth mindset, you can maintain unstoppable momentum on your journey towards personal and professional growth. Embrace the challenges and setbacks you encounter as opportunities to learn and evolve, and you will be well on your way to achieving your goals.

Suggested Reading: "Mindset: The New Psychology of Success" by Carol S. Dweck, Ph.D.

As a Motivational Speaker, I want to inspire you to push past your limitations and continuously evolve. The journey to success is paved with challenges, and your ability to adapt and overcome setbacks will determine your progress. In this section, we will explore additional strategies and tactics to help you maintain momentum and remain steadfast in your pursuit of personal and professional growth.

Creating an Action Plan:

An effective action plan can help you stay focused on your goals and provide a roadmap for overcoming setbacks. Break down your goals into smaller, manageable steps and assign deadlines for each task. By doing so, you create a sense of urgency and motivation to keep moving forward, even in the face of challenges.

Embracing Change:

Change is a natural part of life and personal growth. Instead of resisting change, learn to embrace it and use it as an opportunity to grow and evolve. Stay curious and open-minded, and be willing to explore new ideas and perspectives.

Practicing Mindfulness:

Mindfulness is the practice of being present and fully engaged with your thoughts, emotions, and experiences. By practicing mindfulness, you can develop greater self-awareness and a deeper understanding of your emotions, which can help you manage setbacks more effectively. Regular mindfulness practice, such as meditation or deep breathing exercises, can also reduce stress and improve focus.

Seeking Feedback and Mentorship:

Seeking feedback from others can provide valuable insights into your strengths, weaknesses, and areas for improvement. Engage with mentors, colleagues, or friends who can offer constructive criticism and guidance on your journey towards personal and professional growth. Be open to their input and use it to refine your approach and overcome setbacks.

Maintaining a Positive Attitude:

Maintaining a positive attitude is crucial for overcoming setbacks and maintaining momentum. Focus on the progress you've made and the lessons you've learned along the way. Surround yourself with positive influences and engage in activities that uplift your mood and boost your motivation.

By implementing these strategies and tactics, you can maintain unstoppable momentum and continuously evolve in the face of challenges. Embrace the journey and remain committed to your personal and professional growth, and you will find yourself thriving in any situation.

Suggested Reading: "The Obstacle is the Way: The Timeless Art of Turning Trials into Triumph" by Ryan Holiday.

Example 1: Overcoming a Career Setback

Problem/Question: Jane had been working at a tech company for several years and was expecting a promotion. However, she was passed over for the opportunity, which left her feeling demoralized and questioning her abilities.

Answer/Solution: As an experienced life coach, I advised Jane to focus on the progress she had made in her career thus far and to view the setback as a learning opportunity. Together, we worked on creating an action plan to improve her skills and increase her visibility within the company. With dedication and hard work, Jane was able to secure a promotion during the next evaluation period.

Example 2: Bouncing Back After a Failed Business Venture

Problem/Question: Mike started a small business, but despite his best efforts, it failed within two years. He felt defeated and unsure of whether he should try again or pursue a different path.

Answer/Solution: As an experienced consultant, I worked with Mike to analyze the reasons behind his business's failure and identify areas for improvement. By embracing change and seeking mentorship from successful entrepreneurs, Mike gained new insights and developed a stronger business plan. With renewed determination, he launched a new venture, which has since grown into a thriving business.

Example 3: Struggling with Work-Life Balance

Problem/Question: Sarah was a high-achieving professional juggling a demanding job and family responsibilities. She felt overwhelmed and unable to make progress in either area of her life.

Answer/Solution: As a motivational speaker, I encouraged Sarah to celebrate her accomplishments and practice mindfulness to manage stress more effectively. We also developed a plan to set boundaries at work and allocate more time for self-care and family activities. By implementing these changes, Sarah regained control over her life and achieved a better work-life balance.

Each of these real-life examples demonstrates how individuals can overcome setbacks and maintain momentum on their personal and professional journeys. By embracing lifelong learning, seeking support, and remaining adaptable, anyone can achieve unstoppable momentum and continuously evolve in the face of challenges.

Q: How can I celebrate my progress when I feel like I haven't achieved much?

A: Start by acknowledging even small accomplishments and milestones. Reflect on your journey and recognize the growth you've experienced. Celebrating progress isn't about achieving perfection; it's about recognizing personal development.

Q: How do I overcome a major setback in my career?

A: Reframe the setback as an opportunity to learn and grow. Evaluate your goals, develop an action plan, and seek support from mentors, colleagues, or a life coach to help you bounce back stronger.

Q: How can I maintain momentum when faced with obstacles?

A: Develop resilience by adopting a growth mindset, learning from your mistakes, and embracing change. Keep your goals in mind and surround yourself with positive influences.

Q: How do I maintain a positive attitude during tough times?

A: Practice gratitude, focus on your strengths, and visualize positive outcomes. Engage in activities that bring you joy and surround yourself with supportive people.

Q: What strategies can I use to overcome procrastination and stay on track?

A: Break your goals into manageable tasks, set deadlines, and use time management techniques. Keep yourself accountable by sharing your goals with a trusted friend or coach.

Q: How can I overcome the fear of failure?

A: Accept that failure is a natural part of growth and progress. Reframe failure as a learning opportunity and develop a plan for improvement.

Q: How do I stay motivated when I feel like giving up?

A: Remind yourself of your "why" – the underlying purpose that drives your goals. Surround yourself with positive influences and seek guidance from mentors or coaches when needed.

Q: What's the best way to bounce back from a personal setback?

A: Give yourself time to process your emotions, then develop an action plan to address the issue. Seek support from friends, family, or professionals to help you overcome the setback and move forward.

Q: How can I stay focused on my goals when dealing with personal challenges?

A: Prioritize self-care and ensure your basic needs are met. Create a support network and develop healthy coping mechanisms to manage stress and maintain focus.

Q: How do I regain confidence after a setback?

A: Reflect on your accomplishments, practice self-compassion, and develop a plan to improve your skills. Surround yourself with positive influences and seek feedback from mentors or coaches.

Q: How do I manage setbacks in relationships?

A: Practice open communication, empathy, and actively work on resolving conflicts. Seek guidance from relationship experts or couples therapy if needed.

Q: How can I develop resilience in the face of setbacks?

A: Embrace a growth mindset, practice gratitude, and maintain a strong support network. Focus on your strengths and view setbacks as opportunities for growth.

Q: How do I stay adaptable in a constantly changing world?

A: Embrace lifelong learning, stay informed, and actively seek new experiences. Develop a flexible mindset and be open to change.

Q: How can I maintain momentum when my goals seem unreachable?

A: Break your goals into smaller, manageable tasks, and celebrate each accomplishment. Seek guidance from mentors or coaches to help you stay on track.

Q: How do I deal with setbacks in my health journey?

A: Focus on self-compassion and adapt your goals as needed. Seek support from healthcare professionals and stay committed to your long-term health.

Q: How can I balance celebrating progress with striving for more?

A: Develop a mindset that acknowledges and celebrates growth while maintaining a hunger for continuous improvement. Reflect on your journey, set new

Q: How can I maintain a healthy work-life balance while overcoming setbacks?

A: Prioritize self-care and set boundaries to protect your personal time. Communicate your needs to your employer, and seek support from colleagues, friends, or a life coach to help you navigate challenges.

Q: How can I effectively manage stress during times of setback?

A: Practice mindfulness, meditation, or deep breathing exercises to help manage stress. Engage in physical activity, prioritize sleep, and maintain a healthy diet to support overall well-being.

Q: How do I stay motivated when progress is slow?

A: Focus on the journey rather than the destination. Celebrate small victories, and surround yourself with positive influences. Keep a journal to track your progress and remind yourself of the growth you've experienced.

Q: How can I develop a growth mindset to overcome setbacks and continuously adapt?

A: Embrace the idea that skills and abilities can be improved with effort and practice. View challenges as opportunities for growth, and learn from your mistakes. Seek feedback from others, and maintain a curious, open-minded attitude towards learning and development.

Technique 1: Gratitude Journaling

As a Life Coach, I encourage you to practice gratitude journaling. By listing the things you're grateful for, you can shift your focus to the positive aspects of your life. This practice helps you appreciate your progress and build resilience in overcoming setbacks.

Step 1: Set aside time each day to journal.

Step 2: Write down three things you're grateful for, big or small.

Step 3: Reflect on how these positive aspects have contributed to your progress.

Step 4: Read your gratitude journal entries during tough times to remind yourself of your strengths and achievements.

Technique 2: Visual Goal Setting

Visual goal setting is a powerful tool to keep you motivated and focused on your objectives. Create a vision board or use digital tools to display your goals and aspirations visually.

Step 1: Identify your short-term and long-term goals.

Step 2: Gather images, quotes, or symbols that represent your goals.

Step 3: Arrange these elements on a physical or digital board.

Step 4: Place your vision board in a prominent location or set it as your device wallpaper to keep your goals in sight.

Technique 3: Mindful Meditation

As an Experienced Consultant, I suggest incorporating mindful meditation into your daily routine. This practice helps you manage stress, increase self-awareness, and maintain a clear perspective during setbacks.

Step 1: Find a quiet, comfortable space.

Step 2: Sit or lie down in a relaxed position.

Step 3: Close your eyes and take deep, slow breaths.

Step 4: Focus on your breath, and gently bring your attention back to your breathing whenever your mind wanders.

Step 5: Gradually increase the duration of your meditation sessions as you become more comfortable with the practice.

Technique 4: Seek Constructive Feedback

Continuous growth requires honest evaluation and feedback. Reach out to trusted friends, colleagues, or a mentor to gain valuable insights into your progress and areas for improvement.

Step 1: Identify individuals who can provide constructive feedback.

Step 2: Ask for their honest assessment of your performance and progress.

Step 3: Listen actively and take notes during the conversation.

Step 4: Reflect on the feedback and identify actionable steps for improvement.

By incorporating these techniques into your life, you can celebrate your progress, overcome setbacks, and continuously evolve and adapt on your journey towards unstoppable momentum.

Developing Adaptability and Flexibility

As a Life Coach, I have found that adaptability and flexibility are essential qualities for individuals who want to achieve unstoppable momentum. In a world that is constantly changing, being able to adjust and pivot when necessary is crucial for success. In this section, we'll explore strategies to develop these vital traits, along with real-life examples, questions, and answers to guide you on your journey.

Cultivate a Growth Mindset: Embrace the idea that your abilities can be developed through dedication and effort. This mindset allows you to learn from setbacks and view challenges as opportunities for growth.

Example: Thomas Edison famously failed thousands of times before inventing the light bulb. He viewed each failure as a learning opportunity, which eventually led to his groundbreaking success.

Question: How can you reframe a recent setback as an opportunity for growth?

Answer: Identify the lessons learned from the setback and implement changes to improve future outcomes.

Embrace Change: Accept that change is a natural part of life and view it as an opportunity to learn and grow, rather than something to fear.

Example: When the music industry shifted from physical albums to digital streaming, artists who embraced the change and adapted their business models thrived, while those who resisted struggled.

Question: What is one area of your life where you can embrace change and adapt to new circumstances?

Answer: Identify an area where you are resistant to change and develop a plan to adjust your mindset and actions.

Foster Resilience: Develop the ability to bounce back from adversity and maintain a positive attitude in the face of challenges.

Example: After a devastating knee injury, an athlete could choose to dwell on the setback or focus on their rehabilitation and return to the sport stronger than before.

Question: How can you build resilience in the face of a challenge?

Answer: Develop a support system, practice self-compassion, and set realistic goals for recovery and growth.

Practice Active Listening: Enhance your ability to adapt by actively listening to others and considering diverse perspectives.

Example: A manager who actively listens to employees' concerns and suggestions can make more informed decisions and foster an environment of growth and adaptability.

Question: How can you improve your active listening skills?

Answer: Practice focusing on the speaker, avoiding interruptions, and asking clarifying questions to fully understand their perspective.

Suggested readings for developing adaptability and flexibility:

"Mindset: The New Psychology of Success" by Carol S. Dweck

"Who Moved My Cheese? An Amazing Way to Deal with Change in Your Work and in Your Life" by Spencer Johnson

"Resilience: Hard-Won Wisdom for Living a Better Life" by Eric Greitens

"The Power of Listening: Building Skills for Mission and Ministry" by Lynne M. Baab

By incorporating these strategies into your daily life, you can develop the adaptability and flexibility necessary to continuously evolve and adapt, ultimately achieving unstoppable momentum in your personal and professional life.

Real-Life Story 1: Adapting to a Career Change

Problem/Question: After 15 years in the finance industry, Sarah felt unfulfilled and decided to pursue her passion for health and wellness. However, she was concerned about how to adapt to a new career and find success.

Answer: Sarah started by attending conferences and networking events to learn about the health and wellness industry. She also took courses to deepen her knowledge and skills in her new field. By embracing change and remaining flexible, she was able to adapt to her new career and ultimately opened her own successful wellness center.

Real-Life Story 2: Overcoming a Business Challenge

Problem/Question: Mark owned a small bookstore that faced significant competition from online retailers. He needed to find a way to adapt his business model to stay afloat.

Answer: Mark decided to embrace change and create a unique experience for his customers. He transformed his bookstore into a community hub, offering book clubs, author events, and a cozy cafe. This adaptability allowed his business to thrive in the face of adversity.

Real-Life Story 3: Navigating a Difficult Personal Transition

Problem/Question: After a painful divorce, Michelle felt lost and unsure of how to move forward in her life.

Answer: Michelle sought the support of a Life Coach, who helped her develop resilience and adaptability. She learned to view her situation as an opportunity for personal growth and started engaging in new activities, such as volunteering and joining social clubs. Through embracing change and developing flexibility, Michelle was able to build a fulfilling new life.

Real-Life Story 4: Leading a Team Through Change

Problem/Question: As a manager, Emily faced the challenge of leading her team through a significant organizational restructuring. She needed to find ways to maintain team morale and productivity during this uncertain time.

Answer: Emily focused on fostering open communication and active listening. She held regular meetings to address team members' concerns and provided support and resources to help them adapt to their new roles. By demonstrating adaptability and flexibility herself, Emily was able to successfully guide her team through the transition.

Q: How can I develop adaptability in my personal and professional life?

A: Start by embracing change, seeking out new experiences, and challenging yourself to learn new skills. Cultivate a growth mindset, and be open to feedback and learning from your mistakes.

Q: Why is flexibility important in today's fast-paced world?

A: Flexibility enables you to adapt to new situations and challenges more effectively, which can lead to greater personal and professional success. It also helps you stay resilient during times of uncertainty and change.

Q: How can I become more open to change?

A: Practice being curious and open-minded. Look for opportunities to learn and grow from new experiences, and try to see change as an opportunity rather than a threat.

Q: How can I improve my ability to adapt to new situations at work?

A: Develop strong problem-solving skills, be open to feedback, and learn to collaborate with others. Stay informed about industry trends and changes, and actively seek out opportunities for professional development.

Q: What are some strategies for coping with setbacks and challenges?

A: Focus on what you can control, seek support from others, practice self-compassion, and maintain a positive outlook. Reflect on past experiences and lessons learned, and use those insights to move forward.

Q: How can I become more comfortable with uncertainty?

A: Practice mindfulness and learn to tolerate discomfort. Focus on the present moment and remind yourself that uncertainty is a natural part of life. Develop healthy coping strategies and seek support when needed.

Q: How can I help my team develop adaptability and flexibility?

A: Encourage open communication, collaboration, and innovation. Provide resources and opportunities for professional development, and lead by example by embracing change yourself.

Q: Can adaptability be learned, or is it an innate trait?

A: Adaptability can be developed through practice, experience, and a willingness to embrace change. While some people may have a natural predisposition to be more adaptable, everyone can learn to become more flexible and resilient.

Q: How can I apply adaptability and flexibility to my personal relationships?

A: Practice active listening, empathy, and open communication. Be willing to compromise and adapt to the needs of others, while also maintaining healthy boundaries and self-care.

Q: How do I maintain balance between being adaptable and staying true to my values and goals?

A: Establish clear personal values and goals while remaining open to new experiences and opportunities. Recognize that adaptability doesn't mean abandoning your core beliefs, but rather being flexible in how you pursue and achieve them.

Q: What are some signs that I need to work on my adaptability skills?

A: Difficulty adjusting to change, resistance to new ideas, struggling to learn new skills, and feeling overwhelmed by uncertainty are all signs that you may benefit from developing your adaptability skills.

Q: Can developing adaptability and flexibility help reduce stress?

A: Yes, being adaptable can help you better manage stress by enabling you to navigate challenges and uncertainties more effectively. Cultivating flexibility can lead to increased resilience and a more positive outlook.

Q: How can I practice adaptability in my daily life?

A: Embrace new experiences and challenges, seek out opportunities to learn, and maintain a growth mindset. Practice mindfulness, and be open to feedback and learning from your mistakes.

Q: How can I overcome my fear of change and embrace adaptability?

A: Acknowledge your fear and remind yourself that change is a natural part of life. Practice self-compassion and seek support from others. Focus on the potential benefits and opportunities that change can bring, and start with small steps to build your confidence.

Q: How can I measure my progress in developing adaptability and flexibility?

A: Reflect on your experiences and reactions to change, and notice any improvements in your ability to cope with uncertainty or adapt to new situations. Seek feedback from others, and celebrate your successes along the way.

Q: What are some resources for developing adaptability and flexibility?

A: Books, online courses, workshops, and seminars on topics such as emotional intelligence, resilience, and mindfulness can provide valuable tools and strategies for developing adaptability and flexibility. Networking with others and learning from their experiences can also be helpful.

Q: How can I incorporate adaptability and flexibility into my long-term goals?

A: Set specific, measurable, achievable, relevant, and time-bound (SMART) goals that incorporate adaptability as a key component. Regularly review and adjust your goals as needed to accommodate changes and new opportunities.

Q: How does adaptability contribute to personal and professional growth?

A: Adaptability allows you to seize new opportunities, overcome challenges, and learn from your experiences. It helps you stay relevant in a rapidly changing world and fosters personal and professional development.

Q: How can I encourage others to embrace adaptability and flexibility?

A: Share your experiences and the benefits you've gained from being adaptable. Be a positive role model by demonstrating your own adaptability, and provide support and encouragement to others as they navigate change.

Q: What are some common obstacles to developing adaptability and flexibility, and how can I overcome them?

A: Common obstacles include fear of change, resistance to new ideas, and a fixed mindset. To overcome these barriers, practice mindfulness, cultivate a growth mindset, and seek out opportunities for learning and personal growth.

To develop adaptability and flexibility, follow these practical tools, exercises, and techniques:

Embrace a growth mindset: Cultivate a mindset that embraces challenges, learns from mistakes, and continuously seeks to grow. Recognize that your abilities can be developed through dedication and hard work. This mindset will help you become more adaptable and flexible in various situations.

Mindfulness practice: Regularly practice mindfulness to become more aware of your thoughts and emotions, and learn to respond to change with grace and resilience. This can include meditation, yoga, or deep breathing exercises that help you stay present and focused.

Journaling: Keep a journal to reflect on your experiences, thoughts, and emotions. Use this practice to identify patterns and areas for improvement in your adaptability and flexibility. Write down the challenges you face and the steps you take to overcome them, as well as any successes or lessons learned along the way.

Develop a personal adaptability plan: Create a plan that outlines specific steps you can take to improve your adaptability and flexibility. Break your plan down into smaller, achievable goals, and review your progress regularly to stay motivated and on track.

Role-playing exercises: Participate in role-playing scenarios where you practice adapting to new situations, environments, or roles. This will help you develop the skills necessary to adjust to change quickly and effectively.

Seek diverse experiences: Expose yourself to new ideas, perspectives, and cultures by attending workshops, conferences, or networking events. Engage in activities that challenge your comfort zone and push you to grow.

Learn from others: Seek out mentors, coaches, or colleagues who demonstrate adaptability and flexibility in their lives. Learn from their experiences and ask for advice on how to develop these skills in your own life.

Embrace failure: Recognize that failure is an opportunity for growth and learning. When faced with setbacks, focus on what you can learn from the experience, and use this knowledge to improve your adaptability and flexibility moving forward.

Develop problem-solving skills: Enhance your problem-solving abilities by regularly engaging in activities that challenge your critical thinking and creativity. This can include puzzles, brainstorming sessions, or strategic games.

Practice gratitude: Cultivate an attitude of gratitude by regularly acknowledging and appreciating the positive aspects of your life. This practice can help you maintain a positive outlook and remain adaptable in the face of challenges.

By incorporating these techniques and tools into your daily life, you will be better equipped to develop adaptability and flexibility, allowing you to thrive in an ever-changing world. Remember, as a life coach or motivational speaker, the key to success lies in your ability to continuously evolve and adapt to new situations and challenges.

Staying Inspired and Motivated for Long-Term Success

In this section, we will explore strategies to help you stay inspired and motivated for long-term success. Staying motivated is crucial for maintaining momentum and continuous growth in your personal and professional life. By implementing the following practices, you will be better equipped to stay inspired and achieve long-term success.

Set clear goals: Establish specific, measurable, achievable, relevant, and time-bound (SMART) goals to guide your efforts. Having clear goals will help you stay focused and motivated as you work towards your objectives.

Break goals into smaller tasks: Break your larger goals into smaller, manageable tasks to make progress more achievable.

This will help you maintain motivation as you experience small successes along the way.

Visualize success: Regularly visualize yourself achieving your goals to stay motivated and inspired. This practice can help you maintain a positive mindset and increase your belief in your ability to achieve your objectives.

Surround yourself with positive influences: Connect with individuals who inspire and motivate you. These may be mentors, friends, or colleagues who share your goals and values. Engaging with others who have a positive outlook can help you maintain your own motivation and inspiration.

Develop a strong support network: Build a network of supportive individuals who can provide encouragement and guidance as you work towards your goals. This may include mentors, coaches, or peers who can offer valuable insights and advice.

Embrace lifelong learning: Continuously seek out new knowledge and experiences to stay inspired and engaged. Attend workshops, read books, or take online courses to expand your skillset and stay ahead of the curve in your industry.

Practice self-compassion: Be kind to yourself when you face setbacks or obstacles, and recognize that failure is a natural part of growth. By practicing self-compassion, you can maintain motivation and resilience in the face of challenges.

Celebrate your achievements: Regularly acknowledge and celebrate your accomplishments, no matter how small. This practice will help you maintain motivation and recognize the progress you're making towards your long-term goals.

Monitor your progress: Routinely assess your progress towards your goals and adjust your strategies as needed. This practice will help you stay on track and ensure that you're continuously moving forward.

Create a daily routine: Establish a daily routine that supports your goals and provides structure to your day. Incorporating regular habits, such as exercise, meditation, or journaling, can help you maintain motivation and focus.

Suggested readings:

"The Power of Habit" by Charles Duhigg

"The Miracle Morning" by Hal Elrod

"Drive: The Surprising Truth About What Motivates Us" by Daniel H. Pink

By incorporating these practices into your daily life, you can maintain inspiration and motivation for long-term success. Remember, as a life coach, motivational speaker, or experienced consultant, your ability to stay inspired and motivated is essential for your continued growth and the success of your clients or audience.

Example 1: Overcoming Procrastination

Problem: Sarah, a young entrepreneur, found herself struggling with procrastination. She had big dreams and goals, but she always seemed to put off important tasks and struggled to stay motivated.

Solution: Sarah sought the help of a life coach, who encouraged her to set SMART goals and break them into smaller tasks. The life coach also suggested Sarah create a daily routine and develop a support network to stay accountable. With these strategies in place, Sarah was able to overcome her procrastination, maintain her motivation, and ultimately achieve her goals.

Example 2: Pursuing a Career Change

Problem: John, a 40-year-old accountant, felt unfulfilled in his career and desired a change. However, he was afraid of taking risks and struggled to stay motivated in the face of uncertainty.

Solution: John attended a motivational speaker's seminar and was inspired to take action. He decided to pursue his passion for photography and set clear, achievable goals for his career transition. John surrounded himself with positive influences and embraced lifelong learning to develop his photography skills. As a result, John successfully transitioned into a fulfilling new career and maintained his motivation for long-term success.

Example 3: Bouncing Back from Failure

Problem: Maria, a small business owner, faced several setbacks and failures in her business. She began to lose motivation and questioned her ability to achieve her goals.

Solution: Maria consulted with an experienced business consultant, who advised her to practice self-compassion and embrace failure as a natural part of growth. Maria learned to celebrate her achievements, no matter how small, and regularly visualized herself succeeding. By implementing these strategies, Maria regained her motivation, overcame her setbacks, and achieved long-term success in her business.

These real-life stories showcase the power of staying inspired and motivated for long-term success. By incorporating the strategies discussed in this section, individuals from all walks of life can overcome challenges, maintain momentum, and continuously evolve and adapt on their journey to success.

Q: How can I find inspiration when I feel demotivated?

A: Seek out sources of inspiration, such as books, podcasts, or TED talks by successful individuals who have overcome challenges. Surround yourself with positive and supportive people who encourage your growth and success.

Q: What are some strategies to maintain long-term motivation?

A: Set SMART goals, break them into smaller tasks, develop a daily routine, celebrate small wins, and regularly assess and adjust your progress.

Q: How can I overcome the fear of failure in pursuit of my goals?

A: Embrace failure as a natural part of growth, practice self-compassion, and focus on the lessons learned from setbacks.

Q: What role does a support network play in staying motivated?

A: A strong support network provides encouragement, accountability, and constructive feedback, which can help maintain your motivation during challenging times.

Q: How can visualization techniques improve my motivation?

A: Visualization helps create a clear mental image of your desired outcome, which can increase motivation and focus on achieving your goals.

Q: How can I develop a growth mindset to stay motivated?

A: Embrace lifelong learning, welcome challenges, view setbacks as opportunities for growth, and focus on continuous self-improvement.

Q: How do I stay motivated when faced with obstacles?

A: Keep your end goal in mind, remind yourself of your "why," and develop resilience by learning from setbacks and adapting your approach.

Q: What is the importance of celebrating small wins?

A: Celebrating small wins helps maintain motivation by recognizing progress, boosting self-esteem, and reinforcing positive habits.

Q: How can I stay motivated when I feel overwhelmed by my goals?

A: Break your goals into smaller, achievable tasks, and focus on completing one task at a time. Remember to practice self-compassion and seek support when needed.

Q: How can I use affirmations to boost my motivation?

A: Positive affirmations can help rewire your thought patterns, enhance self-belief, and improve motivation. Repeat affirmations daily to reinforce their impact.

Q: How do I maintain motivation during times of change or uncertainty?

A: Focus on what you can control, practice adaptability, and remain open to new opportunities and learning experiences.

Q: How can I stay motivated when progress is slow or not visible?

A: Track your progress, celebrate small wins, and remind yourself that success often takes time and persistence.

Q: What role does self-awareness play in staying motivated?

A: Self-awareness helps you identify your strengths, weaknesses, and personal values, which can guide your goal-setting process and increase motivation.

Q: How can I avoid burnout while maintaining motivation?

A: Prioritize self-care, set boundaries, and practice effective time management to maintain a healthy balance between work and personal life.

Q: How do I regain motivation after a significant setback?

A: Reflect on the lessons learned, adjust your approach, and set new, achievable goals. Seek support and embrace the opportunity for growth and learning.

Q: How can I maintain motivation when working toward long-term goals?

A: Break your long-term goals into smaller milestones, celebrate your progress, and regularly reassess your goals to ensure they remain relevant and motivating.

Q: How can I stay motivated when I experience self-doubt?

A: Practice self-compassion, focus on your strengths, and surround yourself with positive influences to combat self-doubt and maintain motivation.

Q: What role does passion play in staying motivated?

A: Passion fuels motivation by providing a strong emotional connection to your goals, making it easier to maintain focus and persevere through challenges.

Q: How can I increase my motivation when I'm feeling tired or burnt out?

A: Rest and recharge, engage in activities you enjoy, and practice self-care to replenish your energy levels and reignite your motivation.

Q: How can I maintain motivation when others around me are not supportive or discouraging?

A: Stay focused on your own goals, seek out positive and supportive people, and remind yourself of your "why" to stay motivated in the face of negativity.

Technique 1: Create a Motivation Board

To stay inspired and motivated, create a visual representation of your goals, dreams, and aspirations. Gather images, quotes, and any other items that represent your objectives, and arrange them on a bulletin board or a digital platform. Place your motivation board in a prominent location or set it as your computer or phone background to serve as a constant reminder of your goals.

Technique 2: Practice Gratitude

Cultivate an attitude of gratitude by keeping a daily gratitude journal. Each day, write down at least three things you are grateful for. This practice helps you maintain a positive mindset, focus on the good aspects of your life, and stay motivated to achieve your goals.

Technique 3: Set SMART Goals

Break down your long-term goals into smaller, achievable milestones using the SMART criteria (Specific, Measurable, Achievable, Relevant, Time-bound). This approach makes your goals more manageable and increases your motivation as you experience progress and success.

Technique 4: Engage in Regular Self-Reflection

Take time to reflect on your goals, accomplishments, and setbacks regularly. This process helps you assess your progress, identify areas for improvement, and stay focused on your objectives.

Technique 5: Use Visualization Techniques

Practice visualization by creating mental images of yourself achieving your goals. This mental rehearsal helps program your mind for success and keeps your motivation high.

Technique 6: Find an Accountability Partner

Share your goals with a trusted friend, family member, or mentor who can provide support, encouragement, and guidance. Regular check-ins with your accountability partner can help you stay motivated and on track.

Technique 7: Reward Yourself

Celebrate your achievements and milestones, no matter how small. Plan rewards for yourself as you progress toward your goals. These rewards can serve as an incentive to stay motivated and focused.

Technique 8: Develop a Growth Mindset

Cultivate a growth mindset by embracing challenges, viewing setbacks as opportunities for growth, and believing in your ability to learn and improve. This mindset can help you stay motivated and committed to your long-term success.

Epilogue

As our journey through the world of self-improvement comes to an end, it is essential to remember that personal growth is an ongoing process. The strategies, techniques, and insights shared in this book are not meant to be followed once and forgotten; instead, they should be integrated into our daily lives to help us become better versions of ourselves. Remember, life is a series of moments, and each one presents an opportunity to learn, grow, and evolve.

Afterword

This book has provided you with an extensive roadmap for self-improvement and personal growth. As you continue your journey, we encourage you to explore additional resources, seek support from others, and reflect on your progress regularly. Stay open to change, be patient with yourself, and embrace the challenges that come your way.

Acknowledgments

I would like to express my immense gratitude to all those who have contributed to the creation of this book. From the countless experts and motivational speakers who have inspired me, to my friends and family who have supported me throughout the years, your encouragement and guidance have been invaluable. I am also deeply grateful to my editors, designers, and publishers for their hard work and dedication in bringing this book to life.

www.ingramcontent.com/pod-product-compliance
Lightning Source LLC
Chambersburg PA
CBHW071233070526
44583CB00017B/2169